Studies in Computational Intelligence

Volume 489

Series Editor

J. Kacprzyk, Warsaw, Poland

For further volumes:
http://www.springer.com/series/7092

Moonis Ali · Tibor Bosse
Koen V. Hindriks · Mark Hoogendoorn
Catholijn M. Jonker · Jan Treur
Editors

Contemporary Challenges and Solutions in Applied Artificial Intelligence

 Springer

Editors

Moonis Ali
Department of Computer Science
Texas State University
San Marcos
TX
USA

Tibor Bosse
Agent Systems Research Group
Department of Computer Science
Faculty of Sciences
VU University Amsterdam
Amsterdam
The Netherlands

Koen V. Hindriks
Interactive Intelligence Group
Department of Intelligent Systems
Faculty of Electrical Engineering,
    Mathematics and Computer Science
Delft University of Technology
Delft
The Netherlands

Mark Hoogendoorn
Computational Intelligence Group
Department of Computer Science
Faculty of Sciences
VU University Amsterdam
Amsterdam
The Netherlands

Catholijn M. Jonker
Interactive Intelligence Group
Department of Intelligent Systems
Faculty of Electrical Engineering,
    Mathematics and Computer Science
Delft University of Technology
Delft
The Netherlands

Jan Treur
Agent Systems Research Group
Department of Computer Science
Faculty of Sciences
VU University Amsterdam
Amsterdam
The Netherlands

ISSN 1860-949X   ISSN 1860-9503   (electronic)
ISBN 978-3-319-03275-7   ISBN 978-3-319-00651-2   (eBook)
DOI 10.1007/978-3-319-00651-2
Springer Cham Heidelberg New York Dordrecht London

Preface

Since its origination in the mid-twentieth century, the area of Artificial Intelligence (AI) has undergone a number of developments. While the early interest in AI was mainly triggered by the desire to develop artifacts that show the same intelligent behavior as humans, nowadays the field has become more mature, and researchers have realized that research in AI involves a multitude of separate challenges, besides the traditional goal to replicate human intelligence. In particular, recent history has pointed out that a variety of intelligent computational techniques, part of which are inspired by human intelligence, may be successfully applied to solve all kinds of practical problems. Examples of such intelligent techniques include machine learning, agent technology, and knowledge representation, and examples of application domains include medicine, economics, and incident management, among many others. This sub-area of AI, which has its main emphasis on applications of intelligent systems to solve real-life problems, is currently known under the term Applied Intelligence.

The objective of the International Conference on Industrial, Engineering & Other Applications of Applied Intelligent Systems (IEA/AIE) is to promote and disseminate recent research developments in Applied Intelligence. Held yearly, the conference brings together scientists, engineers and practitioners, who work on designing and developing applications that use intelligent techniques and apply them to a variety of application domains. The current book, which is published in the Studies in Computational Intelligence series by Springer-Verlag, contains short papers authored by participants of the 26th edition of IEA/AIE, which was held in June 2013 in Amsterdam, the Netherlands. This conference was organized by VU University Amsterdam in collaboration with Delft University of Technology and Texas State University-San Marcos, and was sponsored by the International Society of Applied Intelligence (ISAI), Almende B.V., the Benelux Association for Artificial Intelligence, and the Municipality of Amsterdam.

This book discusses contemporary challenges as well as solutions regarding to a variety of aspects related to Applied Intelligence. It is comprised of 30 chapters, which are distributed over 11 parts. The material of each chapter is self-contained and was reviewed by at least two anonymous referees, to assure a high quality.

Readers can select any individual chapter based on their research interests without the need of reading other chapters. We are confident that this book provides useful reference values to researchers and students in the field of Applied Intelligence, enabling them to find opportunities and recognize challenges in the field.

We would like to thank Springer Verlag, and in particular Prof. Janusz Kacprzyk, the editor-in-chief of the Studies in Computational Intelligence series, for providing us the opportunity to publish this volume. Moreover, we thank the senior editor, Dr. Thomas Ditzinger, and the editorial assistant, Holger Schäpe, for their efforts in preparing and publishing this book. We also greatly thank all organizing and program committee members for their hard work in assuring the high quality of this book. And last, but not least, we cordially thank all the authors who made important contributions to the conference and the book. Without their efforts, this book could not have been published.

April 2013   Moonis Ali
                                                                        Tibor Bosse
                                                                 Koen V. Hindriks
                                                               Mark Hoogendoorn
                                                              Catholijn M. Jonker
                                                                        Jan Treur

Contents

Part I: Cognitive Modeling

Modelling Space Perception in Urban Planning: A Cognitive AI-Based Approach 3
Dino Borri, Domenico Camarda

Facilitating Player Interaction in a Dynamic Storytelling Environment 11
Richard Paul, Darryl Charles, Michael McNeill, David McSherry

Part II: Distributed Systems and Networks

Using Agents for Dynamic Components Redeployment and Replication in Distributed Systems 19
Nadim Obeid, Samih Al-Areqi

Prototyping and Evaluation of a Wireless Sensor Network That Aims Easy Installation 27
Takanobu Otsuka, Tatsunosuke Tsuboi, Takayuki Ito

Part III: Evolutionary Algorithms

Winner Determination in Combinatorial Reverse Auctions 35
Shubhashis Kumar Shil, Malek Mouhoub, Samira Sadaoui

Virus Transmission Genetic Algorithm 41
Weixin Ling, Walter D. Potter

Automated Phenotype-Genotype Table Understanding 47
Shifta Ansari, Robert E. Mercer, Peter Rogan

Part IV: Knowledge Representation and Reasoning

An Implementation of a Menu-List Recommendation System
Providing Feedback from User   55
Chika Nishikawa, Akihiko Nagai, Takayuki Ito, Satomi Maruyama

On Representing and Sharing Knowledge in Collaborative Problem
Solving   61
Heba Al-Juaidy, Lina Abu Jaradeh, Duha Qutaishat, Nadim Obeid

Part V: Machine Learning Applications

Constructing Language Models for Spoken Dialogue Systems
from Keyword Set   69
Kazunori Komatani, Shojiro Mori, Satoshi Sato

A Speaker Diarization System with Robust Speaker Localization
and Voice Activity Detection   77
Yangyang Huang, Takuma Otsuka, Hiroshi G. Okuno

A Content Fusion System Based on User Participation Degree
on Microblog   83
Wo-Chen Liu, Meng-Hsuan Fu, Kuan-Rong Lee, Yau-Hwang Kuo

Network Intrusion Detection System Based on Incremental Support
Vector Machine   91
Haiyi Zhang, Yang Yi, Jiansheng Wu

Use of Fuzzy Information for Heterogeneous Performance
Evaluation   97
Mohammad Anisseh, Mohammad Reza Shahraki

Part VI: Optimization

Designing Loss-Aware Fitness Function for GA-Based Algorithmic
Trading   107
Yuya Arai, Ryohei Orihara, Hiroyuki Nakagawa, Yasuyuki Tahara,
Akihiko Ohsuga

Watching Subgraphs to Improve Efficiency in Maximum Clique
Search   115
Pablo San Segundo, Cristobal Tapia, Alvaro Lopez

Decision Making and Optimization for Inspection Planning under
Parametric Uncertainty of Underlying Models   123
Nicholas Nechval, Gundars Berzins, Vadim Danovich,
Konstantin Nechval

Topological Feature Mining for Rambling Activities 131
Masakatsu Ohta, Miyuki Imada

Part VII: Pattern Recognition

Confusion Matrix Based Reweighting 143
Vincent Damian Warmerdam, Zoltán Szlávik

Web Performance Forecasting with Kriging Method 149
Leszek Borzemski, Anna Kamińska-Chuchmała

Part VIII: Problem Solving

Application of the Swarm Intelligence Algorithm for Investigating the
Inverse Continuous Casting Problem 157
Edyta Hetmaniok, Damian Słota, Adam Zielonka

Estimating Mental States of a Depressed Person with Bayesian
Networks 163
Michel C.A. Klein, Gabriele Modena

Multi-objective Optimization Algorithms for Microchannel Heat Sink
Design 169
Ahmed Mohammed Adham, Normah Mohd-Ghazali, Robiah Ahmad

Solution of the Inverse Stefan Problem by Applying the Procedure
Based on the Modified Harmony Search Algorithm 175
Edyta Hetmaniok, Damian Słota, Adam Zielonka, Roman Wituła

Part IX: Robotics

Cascade Safe Formation Control for a Fleet of Underactuated Surface
Vessels Using the DCOP Approach 183
*Alejandro Rozenfeld, Jawhar Ghommam, Rodrigo Picos,
Gerardo Acosta*

UMH's Navigation in Unknown Environment Based on Pre-planning
Guided Fuzzy Reactive Controller 189
Xuzhi Chen, Zhijun Meng, Wei He, Kaipeng Wang

Part X: Special Session on Decision Support for Safety-Related Systems

Developing Context-Free Grammars for Equation Discovery:
An Application in Earthquake Engineering 197
Štefan Markič, Vlado Stankovski

Neural Networks to Select Ultrasonic Data in Non Destructive
Testing 205
Thouraya Merazi Meksen, Malika Boudraa, Bachir Boudraa

Part XI: Special Session on Innovations in Intelligent Computation and Applications

Stairway Detection Based on Extraction of Longest Increasing
Subsequence of Horizontal Edges and Vanishing Point 213
Kaushik Deb, S.M. Towhidul Islam, Kazi Zakia Sultana,
Kang-Hyun Jo

A Heuristic to the Multiple Container Loading Problem with
Preference 219
Tian Tian, Andrew Lim, Wenbin Zhu

Author Index 225

Part I
Cognitive Modeling

Modelling Space Perception in Urban Planning: A Cognitive AI-Based Approach

Dino Borri and Domenico Camarda[*]

Abstract. The study deals with cooperative space conceptualization by humans according to the AI-based cognitive approach and the urban-planning approach of architects and planners. It carries out the diagnosis and the control of example spaces in known urban environments. The paper is oriented toward suggesting system architectures to let spatial agents add structuring degrees to navigated urban spaces and challenge relevant disorientation conditions.

The methodology draws on ontology-based text-mining analysis and statistical interpretation applied to university-class questionnaire surveys, exploring behaviours in human interaction with a space. After an introduction, a case-based discussion of the cooperative conceptualization and representation of space is carried out. The third section shows the ontological results of the case-study, with general results and follow-up discussed in the concluding section.

Keywords: Decision support, Spatial cognition, Environmental planning, Space ontology, Multi agent systems.

1 Research Background

This research explores human agents' approaches to conceptualize and represent a spatial environment during their 'navigation' to reach a spatial objective [1][2].

The problem of orienting and moving in navigation is common in cognitive sciences [3]. Yet, the present work addresses the problem of the role played by

Dino Borri · Domenico Camarda
Technical University of Bari, Italy
e-mail: d.camarda@poliba.it

[*] The present study has been carried out by the authors as a joint research work. Nonetheless, section 1 has been written by D. Borri, sections 2,3,4 have been written by D. Camarda..

M. Ali et al. (Eds.): *Contemporary Challenges & Solutions in Applied AI*, SCI 489, pp. 3–9.
DOI: 10.1007/978-3-319-00651-2_1   © Springer International Publishing Switzerland 2013

diverse space components to support the navigation. A focus is on 'structural' or 'substantial' components, as opposed to apparently 'ornamental' components. An intriguing logical argument about fundamental vs. secondary components in human agents' perception of reality comes out in literature, frequently fostering (particularly in robotics), but also opposing an alleged differentiation between structured or unstructured space navigation [4].

In structured spaces, robotics consider movements and learning easier and more cognitively recognizable than in unstructured spaces. This is different when dealing with human agents. A structured space for a human agent may be represented by an indoor space –such as, e.g., a hospital aisle- that is geometrically simple and empty and, in fact, easy to be navigated with low concentration. Conversely, an outdoor space –e.g., a plaza-, where a human agent moves randomly, in crowded and hardly recognizable geometries, can represent an unstructured space [5].

In particular, we aim at investigating on a potential distinction between 'structure' and 'ornament', in individual and multi-agent perspective in human space perception. The situation is even slightly more complex here than an indoor structured corridor investigated in a previous experimentation [6].

In this framework, the paper carries out the diagnosis and the control of example spaces in known urban environments, with a decision-support system approach for spatial agents' navigation. The methodology draws on ontology-based text-mining analysis and statistical interpretation applied to university-class questionnaire-based surveys, exploring behaviours in human interaction with a space.

2   Spatial Features of a Navigated Environment

In a previous case-study, the navigated space-environment was an indoor university corridor, where students looked for a professor's office [6]. Now, the complexity is raised by a more random and numerous presence of interfering agents in a less mono-dimensional geometry. The questionnaire is shown in figure 1 [7].

A hundred students of a Town & Country planning course of the Politecnico di Bari (Italy) responded to the questionnaire through a university webpage. The physical context was a commercial street, i.e., a complex but geometrically structured urban space. Answer protocols were statistically analysed, looking for primitive space features, namely 'structures' and 'ornaments' [4] and 'landmarks' and 'beacon' (respectively on-way signs or distant guiding target for a navigation task) [8] Text-mining software was applied to protocols, mainly with a keyword-based approach, followed by Principal component analysis (PCA) to reduce variables.

When analysing questionnaire responses, agents' actions show a straightforward target orientation, and a set of elements is common to the two features. PCA was carried out on about 40 keywords for each feature group. The analysis found out numerous variables: the first 5 ones were selected, as they explained about 80% of the total variance, in both cases (figure 2) [6].

The most statistically significant features of the lists roughly form mutually exclusive clusters, i.e., architectural (in ornaments) and direction/position (in structures) elements. They are clearer than in the indoor experiment [6].

When analysing protocols for 'landmark' or 'beacon' features, again 5 aggregate components found by PCA explained approximately 80% of the total variance [7][8]. However, in this case, sets are only roughly logically distinguishable and mutually exclusive.

Situation 'A': You are in Bari, at the crossroads between via Sparano and via Nicolai, and decide to go north, to a shop in via Sparano, crossing the space in between.
Question A1: Find out the shop you want to go and declare the reasons for your visit.
Question A2: Describe the actions you carry out to reach the shop of your interest
Question A3: Describe in detail the "<u>substantial elements</u>" of the space in which you move, being of help or obstacle in your reaching the shop of your interest ("*substantial elements*" are intended as *spatial elements and their physical qualities, or substances such as materials, dimensions, physical barriers/helpers etc...*)
Question A4: Describe in detail the "<u>ornamental elements</u>" of the space in which you move, conditioning the actions in your reaching the shop of your interest ("*ornamental elements*" are intended as *objects, shapes, colours, lights, aesthetics etc...*)
Question A5: Describe in detail your general sensations and preferences concerning the "<u>substantial elements</u>" and the "<u>ornamental elements</u>" of the space in which you move.

Situation 'B': You are in Bari, at the crossroads between via Sparano and via Nicolai, and decide to go north, to the Art Desco bar in via Sparano, crossing the space in between.
Question B1: Describe the actions you carry out to reach Art Desco bar.
Question B2: Describe in detail the "<u>substantial elements</u>" of the space in which you move, being of help or obstacle in your reaching Art Desco bar.
Question B3: Describe in detail the "<u>ornamental elements</u>" of the space in which you move, conditioning the actions in your reaching Art Desco bar.
Question B4: Describe in detail your general sensations and preferences concerning the "<u>substantial elements</u>" and the "<u>ornamental elements</u>" of the space in which you move.

Fig. 1 The questionnaire survey

1. Structural features	2. Ornamental features
Buildings and service areas	Physical elements at urban scale
Elements of position and directions	Phys. elements at architectural
Shopping-related elements	scale
Moving aids and supports	Elements of shopping and
Elements of parking & vehicular	leisure
mobility	Elements of street furnishing
	Tourist-oriented elements

Fig. 2 PCA on structures and ornaments

As a whole, the robotics division between structures and ornaments in navigation tasks seems to be challenged by a more complex and fuzzy situation, somehow confirming Goodman's scepticism [4]. This does not necessarily mean that spatial features are loosely described, because in real urban contexts they belong to coherent ontologies, useful for decision support systems. Also, agents' perceptions are distorted by the fact that an urban street is not a perfect mono-dimensional space, and transversal navigations and distractions play important roles. This induces a low structuring degree that is difficult to be analysed [9]. Further, the intersection of statistical results depends on an actual multi-agent, rather than single-agent, process where students often strolled around in groups, delivering similar survey responses [10]. In this complex condition, the imperfect result aggregation achieved is clearly consistent with a cognitive process not completely understood.

3   Ontological Results

The above discussion confirms that spacescape is hardly identifiable as a single, static and a-priori snapshot for a whole community, as popularly believed. It varies among agents and with time, with features that are not always so obvious as alleged [11].

As mentioned, this paper shows the most recent stage of a research carried out in different contexts. In the primary set of experimentations carried out in Mediterranean countries [12], spaces were conceptualized by agents as made up of different entities, diverse and multidirectional trajectories with arcs mutually interconnected geometrically, as well as functionally and perceptively. As mentioned in logical and topological literature [13][14] such space is an ill-structured entity overall, whose structuring degree, for representation or navigation aims, strongly depends on a number of quali-quantitative features. That research tried to single out some of them, drawing on group-reasoning-based, multi-agent perceptions and drawing out cooperative conceptualizations of space. The present research has allowed a sort of fine-tuning of some features, decreasing the navigation scale. The space-environment analysed is clearly a structured (although not completely) [14] space, especially when compared to an urban region. An attempt to integrate overall results ontologically was then carried out. The ontology starts from a simple concept tree, where the concepts in the two researches are merged into classes and linked by logical properties [15].

The ontology is written in OWL 2.1, so allowing the processing of human-generated information through software agents rather than merely by human agents. The Space environment ontology can be then suitable to be processed by reasoning and query engines, e.g. for decision support (figure 3) [15].

Of course, the above ontological description is a simplified conceptualization, just aimed at exploring possible models of representation of space perception able to couple with such complexity and to preserve embedded pieces of information. Therefore, it should be regarded as an explorative, rather than deliberative, system approach to spatial knowledge. Fortunately, under a planning or decisionmaking perspective, this complex structure of space perception is increasingly considered by DSSs as a richer framework to allow more aware decisions. Therefore, DSS architectures have started embedding approaches and methodologies, in order to grasp such multifaceted representation of spatial perceptions. Integrated attempts have been difficult and with mixed results, to date, in that they still fail to provide a rich but manageable synthesis to decisionmakers in need of support [16].

4   Concluding Remarks

The present paper deals with the cooperative conceptualization and representation of space, aiming at supporting decision and navigation in space environments. As a main outcome, investigated space features seem to somehow affect space conceptualization and navigation.

The structure/ornament dichotomy emerges rather evidently from the analysis, even if some 'grey areas' stay between structures and ornaments, characterizing the basic fuzziness of the two features. Even the increased degree of uncertainty from a quasi-mono-dimensional situation (an indoor corridor) to a quasi-bi-dimensional situation (an urban street) does not substantially compromise that dichotomous clustering of features.

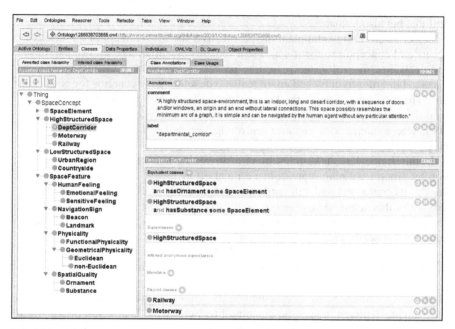

Fig. 3 Map of the space-environment ontology (Protégé software, excerpt)

Also, ornaments do arguably play primary roles in reaching the structure searched for (the shop) and enhancing the structuring degree (usability) of the urban street to accomplish the task. Admittedly, the degree of structuring space ensured by orna-ments proves to be weaker in an urban street than in a indoor corridor context, because of that weaker mono-dimensionality and stronger space complexity. Yet a structuring degree does exist, being produced by the deliberate   intentionality of agents during their navigation tasks guided by their target-reaching aim, confirmed by their search for features of a facilitated path. This might suggest that intentionality seems to actually create structuring degrees.

However, such agents' intentionality is fostered by agents' ability to identify and associate spatial characteristics (i.e., to 'structure' space) and support their navigation in a lowly structured environment [17]. In this perspective, association ability may represent an actual creative attitude of agents, rather critical in the navigation effort. This is an intriguing argument, able to transpose the traditionally physical and geometrical concept of space structuring to a more primitive and abstract approach level, perhaps understated by recent mainstream, application-oriented literature but greatly present in classical speculation [18][19].

In sum, the perceivable but not completely clear characterization of space features, as well as the non-negligible importance of creativity connected to space structuring efforts in space navigation do represent a challenge for the logical argument (particularly in robotics) about fundamental vs. secondary components in agents' perception of space. This should be taken in due consideration when modelling space for navigation and –in general- spatial decision support systems, particularly within a multi-agent-based approach in which agents are not only artificial but also human.

Under an environmental planning perspective, the involvement of a plurality of agents and conceptions about space still remains essential. The knowledge raised and exchanged in real group-reasoning processes represents a critical issue to be managed in order to draw and attain realistic development scenarios for communities. The re-composition of the knowledge–action dichotomy in spatial planning, long discussed by literature [e.g.: 20], can find a catalytic layout in multi-agent DSS, where interactions enhance knowledge-intensive and action-oriented planning processes.

In this framework, research aiming at involving and improving spatial cognition by group reasoning represents a critical effort toward effective management, besides its contingent and still partly unresolved difficulties. The research is rather explicit in this sense, sketching out multiform spatial perceptions and consequent categorizations, in turn involving multiform possibilities for DSS in enhancing development scenarios.

The possibility of singling out logical *IF-THEN* rules, for example, is an intriguing horizon, particularly useful for the drawing out of navigation-support layouts for multi-agent spatial objectives.

However, the cognitive interactions of the agents in reflexive-creative tasks, as for example depicted in specific planning literature [e.g.: 21], would benefit from system architectures based on spatial ontologies. Far from being depressed in narrow patterns, professional creativity would in fact focus its creative effort on organizing originally and effectively spatial concepts and primitives from ontologic libraries and database previously collected. In this complexity-intensive context, building up a system architecture based on spatial ontologies could have a fair motivation in environmental planning. Therefore, it is an interesting perspective for future research.

References

1. Garling, T., Evans, G.W.: Environment, Cognition, and Action: An Integrated Approach. Oxford University Press, New York (1991)
2. Ferber, J.: Multi-Agent Systems: An Introduction to Distributed Artificial Intelligence. Addison-Wesley, London (1999)
3. Tversky, B., Hard, B.M.: Embodied and disembodied cognition: Spatial perspective-taking. Cognition 110, 124–129 (2009)
4. Goodman, N.: The Structure of Appearance. Harvard UP, Cambridge (1951)

5. Hirtle, S.C.: Neighborhoods and landmarks. In: Duckham, M., Goodchild, M.F., Worboys, M.F. (eds.) Foundations of Geographic Information Science, pp. 191–230. Taylor & Francis, London (2003)
6. Borri, D., Camarda, D.: The cooperative conceptualization of urban spaces in AI-assisted environmental planning. In: Luo, Y. (ed.) CDVE 2009. LNCS, vol. 5738, pp. 197–207. Springer, Heidelberg (2009)
7. Borri, D., Camarda, D., Stufano, R.: Memory and creativity in cooperative vs. non cooperative spatial planning and architecture. In: Luo, Y. (ed.) CDVE 2010. LNCS, vol. 6240, pp. 56–65. Springer, Heidelberg (2010)
8. Kelly, D.M., Bischof, W.F.: Orienting in virtual environments: How are surface features and environmental geometry weighted in an orientation task? Cognition 109, 89–104 (2008)
9. Danziger, D., Rafal, R.: The effect of visual signals on spatial decision making. Cognition 110, 182–197 (2009)
10. Wooldridge, M.: An Introduction to Multi-Agent Systems. Wiley, London (2002)
11. Day, S.B., Bartels, D.M.: Representation over time: The effects of temporal distance on similarity. Cognition 106, 1504–1513 (2008)
12. Khakee, A., Barbanente, A., Camarda, D., Puglisi, M.: With or without? Comparative study of preparing participatory scenarios using computer-aided and traditional brainstorming. Journal of Future Research 6, 45–64 (2002)
13. Simon, H.A.: The Sciences of the Artificial. MIT Press, Cambridge (1969)
14. Shirali, S., Vasudeva, H.: Metric Spaces. Springer, London (2005)
15. Borri, D., Camarda, D.: Spatial ontologies in multi-agent environmental planning. In: Yearwood, J., Stranieri, A. (eds.) Technologies for Supporting Reasoning Communities and Collaborative Decision Making: Cooperative Approaches, pp. 272–295. IGI Global Information Science, Hershey Pa (2010)
16. Borri, D., Camarda, D.: Visualizing space-based interactions among distributed agents: Environmental planning at the inner-city scale. In: Luo, Y. (ed.) CDVE 2006. LNCS, vol. 4101, pp. 182–191. Springer, Heidelberg (2006)
17. Barkowsky, T., Knauff, M., Ligozat, G., Montello, D.R.: Spatial Cognition: Reasoning, Action, Interaction. Springer, Berlin (2007)
18. Shelton, A.L., McNamara, T.P.: Systems of Spatial Reference in Human Memory. Cognitive Psychology 43, 274–310 (2001)
19. Bostock, D.: Space, Time, Matter, and Form: Essays on Aristotle's Physics. OUP, Oxford (2006)
20. Friedmann, J.: Planning in the Public Domain: From Knowledge to Action. Princeton University Press, Princeton (1987)
21. Schön, D.A.: The Reflexive Practitioner. Basic Books, New York (1983)

Facilitating Player Interaction in a Dynamic Storytelling Environment

Richard Paul, Darryl Charles, Michael McNeill, and David McSherry

Abstract. Enabling players to interact with stories generated using artificial intelligence planning techniques, and thus exert their own influence on the emergent narrative, is an important challenge in the development of interactive computer game worlds. We focus in this paper on story planning in a dynamic storytelling environment and the problem that arises when plan steps are reduced to primitive actions that can be executed in the game world, but lack sufficient context for players to understand their purpose from a narrative perspective. We propose a solution to this problem in which story plans are represented at two different levels of abstraction, one that allows for meaningful player interaction, and another that enables plan steps to be executed in the virtual world.

1   Introduction

Increasingly, commercial computer games are being designed to give players more control over game narrative by integrating highly emotive, interactive story elements into the gameplay. Single-player role-playing games (RPGs) are traditionally rich in story design, allowing players to explore freely and engage with stories that interest them, although variations in story content and structure are usually predetermined rather than being dynamic and adaptive. Multiplayer RPGs have an additional focus on social interaction but are limited in their ability to create engaging stories and player actions may have limited impact on the stories that other players experience [1, 2].

Enabling the creation of more dynamic and engaging storylines has become a major focus of research interest in the Interactive Digital Storytelling (IDS)

Richard Paul · Darryl Charles · Michael McNeill · David McSherry
School of Computing and Information Engineering, University of Ulster,
Coleraine BT52 1SA, Northern Ireland, UK
e-mail: rjs.paul@btinternet.com, mdjmcneill@gmail.com,
        {dk.charles,dmg.mcsherry}@ulster.ac.uk

M. Ali et al. (Eds.): *Contemporary Challenges & Solutions in Applied AI*, SCI 489, pp. 11–16.
DOI: 10.1007/978-3-319-00651-2_2   © Springer International Publishing Switzerland 2013

community. Examples of IDS systems with a particular emphasis on interactive drama include Anchorhead [3], PaSSAGE [4], Mimesis [5], and Façade [6]. In common with I-Storytelling [7], Opiate [8] and Haunt 2 [9], the emphasis of our own recent work on the MIST system [10] is on story planning in dynamic storytelling environments. Hierarchical task network (HTN) planning [11] plays a key role in our approach by enabling the creation of complex, non-linear stories that would otherwise not be feasible due to the authorial load demanded by prescripting. As we have shown in previous work, both player and non-player character (NPC) actions can also be allowed to have significant impact on an ongoing story by combining HTN planning with techniques for repairing an incomplete story when plan steps are invalidated by the actions of autonomous characters in the game world [10].

In this paper, we present techniques for enabling players to interact with ongoing stories in MIST, and thus exert their own influence on the emergent narrative. In Section 2, we describe how player interaction is facilitated in our approach by allowing players to select from alternative plan steps at appropriate points in an ongoing story. We also show how the decomposition of plan steps to the level of primitive actions can result in a loss of story context that makes it difficult for players to understand their purpose from a narrative perspective. In Section 3, we describe how this problem is addressed in our approach by a mechanism for dual representation of story plans. Our conclusions are presented in Section 4.

2   Interactive Storytelling in MIST

Artificial intelligence (AI) planning techniques often play an important role in storytelling systems. For example, Mimesis [5] and GADIN [12] use AI planning algorithms in which story actions are defined in terms of STRIPS operators [13] for achieving a set of goals. In common with I-Storytelling [10], our approach to storytelling in MIST is based on HTN planning [11], in which an overall task to be performed is given as input rather than a set of goals to be achieved, and planning knowledge is represented as a collection of methods for task decomposition rather than STRIPS operators. The HTN planner uses the methods provided by the story author to break the overall task down into a set of smaller sub-tasks. The sub-tasks identified are then further broken down, if necessary, to the level of primitive actions that can be executed as plan steps to perform the overall task.

The Story Manager. The storytelling process is coordinated in MIST by a component called the story manager whose role is analogous to that of the dungeon master in the traditional tabletop role-playing game Dungeons and Dragons [14]. The story manager uses the story planning methods provided by the story author to generate a story instance that can be applied in the current world state and initiates the story's execution by assigning story roles to NPCs and player characters in the game world. As the story plan is executed in the game world, the story manager assigns tasks to NPCs and player characters according to their roles in the story. It also monitors the story's progress and attempts to repair the story if it detects that one or more steps in the story plan have been invalidated as a result of changes in the world state.

Story Generation. The starting point for story generation in MIST is a *story headline*, which may include variables representing roles to be assigned to game characters. Other variables in a story headline, or introduced in the planning process, may represent items or locations in the game world. For example, the story headline for a theft and recovery story might be:

I is stolen from V by T and recovered by H

In this example, a valuable item (I) is stolen from a victim (V) by a thief (T) and recovered by a hero (H). Table 1 shows a possible decomposition of the story headline. The role of thief (T) might be assigned to a player character called Bob and the roles of victim (V) and hero (H) to NPCs called Mary and Sam. Actual steps in the example story might then include:

Bob takes Magic Ring from Mary   Mary asks Sam to recover Magic Ring

Player Interaction. The planner's ability to generate multiple story plans for a given world state provides the basis of our approach to facilitating player interaction, which currently assumes that only a single player character is involved in a given story. At each point in the story requiring some action to be performed by the player, the story manager uses the generated plans to identify all possible actions that could be performed by the player while enabling the story to continue in a way that is consistent with at least one of the generated plans. It then asks the player to select a preferred action from these alternative plan steps, and the story continues with the action that the player selects.

Loss of Story Context. A limitation of the above approach to facilitating player interaction is that the decomposition of plan steps to primitive actions in the traditional HTN formalism may result in a loss of story context. It may thus be difficult for players to understand the purpose of actions from which they are asked to select. For example, suppose all the generated plans begin with Bob getting a weapon with which to ambush and rob the victim. There may be several ways for Bob to obtain a weapon, such as going to a shop to buy a sword, or going to another location (e.g., a cave) to retrieve a hidden weapon. With story plans reduced to primitive actions that can be executed in the game world, and Bob initially at Lake 3, the alternatives available to Bob as his first plan step might include:

Bob walks from Lake 3 to Shop 1   Bob walks from Lake 3 to Cave 2

The problem is that the purpose of these alternative actions may not be clear to the player. For example, why should Bob go to Cave 2? In Section 3, we present a solution to this problem in which the traditional HTN formalism is extended to enable plan steps to be represented at a level of abstraction that preserves the context needed for players to make informed choices between alternative actions in an ongoing story.

Table 1 Example HTN decomposition of a story headline in which an item (*I*) is stolen from a victim (*V*) and recovered from the thief (*T*) by a hero (*H*)

Story Headline	Initial Story Decomposition	Story Plan
I is stolen from *V* by *T* and recovered by *H*	*T* gets a weapon *W*	*T* walks from *L1* to *L2* *T* picks up *W* at *L2*
	T ambushes *V*	*T* walks from *L2* to *L3* *T* waits for *V* at *L3* *V* walks from *L4* to *L3* *T* attacks *V* with *W*
	T takes *I* from *V*	*T* takes *I* from *V*
	T escapes with *I*	*T* walks from *L3* to *L5*
	V seeks help from *H*	*V* walks from *L3* to *L6* *V* asks *H* to recover *I*
	H agrees to help *V*	*H* agrees to help *V*
	H recovers *I* from *T*	*H* walks from *L6* to *L5* *H* takes *I* from *T* *H* walks from *L5* to *L6* *H* gives *I* to *V*

3   Dual Representation of Story Plans

Our proposed solution to the problem highlighted in Section 2 extends the traditional HTN formalism to enable story authors to create two complementary HTNs (HTN-H and HTN-L) that represent plan steps at the higher (H) and lower (L) levels needed for players to make informed choices between alternative actions and for plan steps to be executed in the game world. In the extended HTN formalism, the story author provides HTN-H methods for decomposing the story headline into plan steps that preserve the context necessary for players to understand their purpose e.g.,

T goes to *L2* to pick up *W*

Where necessary, the author also provides HTN-L methods for decomposing the HTN-H plan steps into primitive actions to be executed in the game world e.g.,

T walks from *L1* to *L2*   *T* picks up *W* at *L2*

In the storytelling process that we now describe, plan steps generated from HTN-H, the higher level HTN, are used to narrate story events to the player and also to present alternative plan steps to the player at appropriate points in the narrative.

Interactive Storytelling Process. First the HTN-H methods provided by the story author are used to generate all possible stories that can be applied in the current world state. A tree representation of the HTN-H plans is then traversed by the story manager while interacting with the player at points in the ongoing story where a choice of plan steps is available to the player character. Each path from the root node to a leaf node in the plan tree represents a complete story plan at the higher level of abstraction. If the next story step is one that is performed by the player character, then the story manager presents the alternative actions (i.e., HTN-H plan steps) available to the player at this point in the story. The story manager then uses the HTN-L methods provided by the author to decompose the HTN-H plan step selected by the player to the level of primitive actions that can be executed in the game world. When the HTN-L actions for the selected HTN-H plan step have been executed, the story continues with the action selected by the player now as the current node in the HTN-H plan tree. In the case of a story action performed by an NPC, one of the alternative actions (i.e., HTN-H plan steps) is selected at random, and any HTN-L methods provided by the author for its decomposition are applied with no need for player interaction.

Theft and Recovery Example. For the theft and recovery story discussed in Section 2, Fig. 1 shows part of the plan tree that might be created by the story manager from the set of HTN-H plans that can be applied in the current world state. With the thief role again assigned to a player character called Bob, the plan steps selected by the player at each stage of the ongoing story are highlighted in bold. Possible HTN-L decompositions of the selected plan steps are also shown in the figure. With plan steps from which the player is asked to choose represented at the higher level of abstraction (HTN-H) in the extended HTN formalism, the player should now have no difficulty in understanding their purpose in the context of the theft and recovery story.

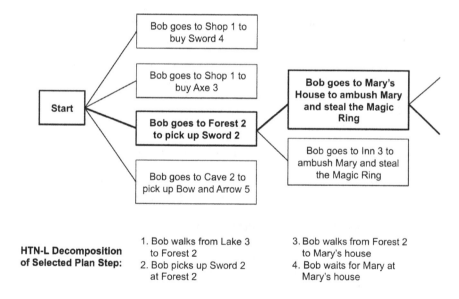

Fig. 1 Example player choices at the beginning of a theft and recovery story

4   Conclusions

We presented an approach to facilitating player interaction with ongoing stories in a dynamic storytelling environment (MIST), thus enabling players to exert their own influence on the emerging narrative. We also demonstrated the loss of story context that may occur when plan steps are decomposed to the level of primitive actions, making it difficult for players to understand the purpose of alternative plan steps. The dual representation of story plans enabled by our extension of the traditional HTN formalism provides a solution to this problem in which the additional authoring load is compensated by the ability to communicate story content to players more effectively.

References

1. Achterbosch, L., Pierce, R., Simmons, G.: Massively Multiplayer Online Role-playing Games: the Past, Present, and Future. ACM Comput. Entertain. 5 (2008)
2. Tychsen, A., Hitchens, M.: Ghost Worlds – Time and Consequence in MMORPGs. In: Göbel, S., Malkewitz, R., Iurgel, I. (eds.) TIDSE 2006. LNCS (LNAI), vol. 4326, pp. 300–311. Springer, Heidelberg (2006)
3. Nelson, M., Roberts, D., Isbell, C., Mateas, M.: Reinforcement Learning for Declarative Optimization-based Drama Management. In: AAMAS 2006, pp. 775–782. ACM, New York (2006)
4. Thue, D., Bulitko, V., Spetch, M., Wasylishen, E.: Interactive Storytelling: A Player Modelling Approach. In: Schaeffer, J., Mateas, M. (eds.) AIIDE 2007, pp. 43–48. AAAI Press (2007)
5. Young, R.M., Riedl, M.O., Branly, M., Jhala, A., Martin, R.J., Saretto, C.J.: An Architecture for Integrating Plan-based Behavior Generation with Interactive Game Environments. Journal of Game Development 1, 51–70 (2004)
6. Mateas, M., Stern, A.: Façade: An Experiment in Building a Fully-Realized Interactive Drama. In: Game Developer's Conference, Game Design Track (2003)
7. Cavazza, M., Charles, F., Mead, S.: Character-based Interactive Storytelling. IEEE Intell. Syst. 17, 17–24 (2002)
8. Fairclough, C.: Story Games and the OPIATE System. PhD Thesis, Trinity College Dublin (2004)
9. Magerko, B.: Story Representation and Interactive Drama. In: Young, R.M., Laird, J. (eds.) Artificial Intelligence and Interactive Digital Entertainment Conference, pp. 87–92. AAAI Press (2005)
10. Paul, R., Charles, D., McNeill, M., McSherry, D.: Adaptive Storytelling and Story Repair in a Dynamic Environment. In: André, E. (ed.) ICIDS 2011. LNCS, vol. 7069, pp. 128–139. Springer, Heidelberg (2011)
11. Erol, K.: Hierarchical Task Network Planning: Formalization, Analysis, and Implementation. PhD Thesis, University of Maryland (1996)
12. Barber, H.: Generator of Adaptive Dilemma-based Interactive Narratives. PhD Thesis, University of York (2008)
13. Fikes, R.E., Nilsson, N.J.: STRIPS: A New Approach to the Application of Theorem Proving to Problem Solving. Artif. Intell. 2, 189–208 (1971)
14. Dungeons and Dragons. TSR (1977)

Part II
Distributed Systems and Networks

Using Agents for Dynamic Components Redeployment and Replication in Distributed Systems

Nadim Obeid and Samih Al-Areqi

Abstract. Availability is one of the important criteria that affect the usefulness and efficiency of a distributed system. It mainly depends on how the components are deployed on the available hosts. In this paper, we present a generic agent-based monitor approach that supports the dynamic component redeployment and replication mechanisms which were presented in Avala and E-Avala. Avala and E-Avala were proposed to improve availability in large and distributed component-based systems via redeployment and replication. By reifying the interaction between the system and components, agents can detect when it is necessary to change the configuration and whether redeployment or replication is more appropriate.

Keywords: Distributed Systems, Agents, Availability, Redeployment.

1   Introduction

Distributed Systems (DS) have to face the problem of disconnected operations. In addition to the fact that the initial deployment architecture may not be very suitable, it is difficult to predict, at design time, the applications which the DS has to deal with. Therefore, finding and maintaining a desirable (e.g. availability) deployment architecture that satisfies a given set of constraints is a challenging problem. This is due to the facts that (1) there are many parameters which influence the selection of an appropriate deployment architecture   (2) the space of possible architectures is large and (3) there may be constant need to change locations of components to meet changing requirements. This leads to some problems such as availability, dependency management [15]], and dynamic configuration. Hence,

Nadim Obeid · Samih Al-Areqi
Department of Computer Information Systems,
King Abdullah II School for Information Technology, The University of Jordan
e-mail: obein@ju.edu.jo

M. Ali et al. (Eds.): *Contemporary Challenges & Solutions in Applied AI*, SCI 489, pp. 19–25.
DOI: 10.1007/978-3-319-00651-2_3   © Springer International Publishing Switzerland 2013

mechanisms such as components replication and redeployment may be necessary in order to improve availability and reliability [1, 2, 10, 12, 5].

In this paper, we present a generic agent-based monitor approach that supports the dynamic component redeployment and replication mechanisms which were presented in Avala [12] and E-Avala [3]. E-Avala improves on Avala by (1) considering positive and negative dependencies among components and (2) implementing replication taking into consideration negative dependencies. By reifying the interaction between the system and application components, agents can detect when it is necessary to change the configuration and whether redeployment or replication is more appropriate.

In section 2 we discuss Avala and E-Avala. In section 3 we present the agent-based redeployment approach. In Section 4 we discuss previous approaches to replication and redeployment.

2   Avala and E-Avala

In this section we give a brief presentation of Avala [6] and E-Avala [2]. Let h_1, h_2, ..., h_k ($1 \leq k$) stands for hosts, $MEM(h_i)$ be the memory of h_i. C_1, ..., C_n ($1 \leq n$) stands for components, $MEM(C_i)$ be the memory of Ci and $FREQ(Ci, Cj)$ be the frequency between components C_i and C_j. The Avala algorithm [6] starts by ranking all hardware nodes and software components as follows:

$$\text{IHR}_i = \sum_{j=1}^{k} REL(h_i, h_j) + MEM(h_{i)})$$ (1)

The ranking of software components is performed as follows:

$$\text{ICRi} = d * \sum_{j=1}^{n} FREQ(C_i, C_j) + \frac{E}{MEM(C_i)}$$ (2)

Where d denotes contributions of host memory and E contributions of event size of interactions between C_i and C_j.

The next software component to be assigned to h, is the one with the smallest memory requirement and which would contribute maximally to the availability function if placed on h. The Component Rank (CR) is calculated as follows:

$$CR(Ci, h) = D_1(Ci, h,n) + D_2(Ci, h)$$ (3)

where $D_1(C_i, h, n) = d * \sum_{j=1}^{n} FREQ(C_i, MC_j) * REL(h, f(MC_j))$

and $D_2(C_i, h, n) = \frac{E}{MEM(C_i)}$

and MC_j is a shorthand for mapped Cj, $f(MC_j)$ is a function that determines the hosts of mapped components, $REL(h, f(MC_j))$ is a function that determines the reliability between selected host h, and hosts of mapped components.

Host Rank (HR) is calculated as follows:

$$HR(h_i) = \sum_{j=1}^{m} REL(h_i, MH(h_j)) + MEM(h_i)$$ (4)

where m is number of hosts that are already selected.

E-Avala [2] employs the notion Depend(C_i, C_j), not present in Avala, as follows:

$$Depend(C_i, C_j) = \begin{cases} 1 & \text{if Ci depends on Cj} \\ -1 & \text{if Cj depends on Ci} \end{cases} \quad (5)$$

Furthermore, E-Avala takes into considration whether or not is a need for data consistancy check regarding a C_i as shown below in (6):

$$Consis(C_i) = \begin{cases} 1 & \text{if Ci does requires data consistency} \\ 0 & \text{Otherwise} \end{cases} \quad (6)$$

Let h be the selected host, l is the level of dependency for system configuration, determined by the designer, and *nm* be number of mapped components (i.e., already been assigned to selected hosts), E-Avala uses the same equations of Avala to calculate the intial ranking and distribution. It improves on Avala by employing two additional functions: RCR (resp. Consis-RCR) that compute Replicate Component Rank without (resp. with) consideration for data consistency.

$$RCR(C_i, h, n, nm) = D_3(C_i, h, n) + D_1((C_i, h, nm) \quad (7)$$

where $D_3(C_i, h, n) = \sum_{p=1}^{n} Depend(C_p, C_i) + \dfrac{l + 2E}{MEM(C_i)}$

$$Consis\text{-}RCR(C_i, h, n, nm) = D_3(C_i, h, n)*(1\text{-}Consis(C_i) + D_1(C_i, h, nm) \quad (8)$$

E-Avala makes a comparison between the selected components for redeployment determined by CR (cf. (3)) and those to be replicated determined by RCR (cf. (5)). The selected component will be the one with the highest value of CR and RCR and that satisfies the constraints of memory, *Loc*, and *Colloc* with respect to the current host h and components which are already assigned. This process is be repeated until h is saturated. The performance of Avala and E-Avala is discussed in [2].

3 Agent-Based Redeployment

Agents are specialized autonomous problem solving entities that are suitable for problem solving in DS [6, 8, 9, 10, 11, 12]. The use of agents enables us (1) to keep track of the communication cost, (2) to mange dynamic reconfiguration while the system is operational and (3) to choose the better mechanism (e.g., redeployment or repplication) to maintain availability at minimal cost.

Let H_R (resp. H_T) stands for the host of the requesting component, C_R, (resp. target component C_T). We employ two kinds of Agents: (1) Comp-Agent (CPA), which has the required information about its host's components and has the ability to monitor any frequent interactions between a component on its host and components on other hosts and (2) Comm-Agent (CMA), which manages the communications with the other Host's *CMAs*. Let $Cor(C_R, C_T)$ stand for the cost of request between C_R and C_T. When $Cor(C_R, C_T)$ becomes high (e.g., above a certain threshold), CPA of H_T will negotiate with the CPA of H_R (through CMAs of H_T and H_R) in order to agree on one of the following options : (1) redeploying C_R in

H_T, (2) redeploying C_T in H_R, (3) replicating C_R in H_T, (4) replicating C_T in H_R or (5) no change. Assuming $H_R \neq H_T$, $Cor(C_R, C_T)$ can be defined as follows:

$$Cor(C_R, C_T) = freq(C_R, C_T)*eventsize(C_R, C_T)/reliable(H_R, H_T) \qquad (9)$$

where $freq(C_R, C_T)$ represents the frequency of interaction between C_R and C_T, $eventsize(C_R, C_T)$ denotes the size of interactions between C_R and C_T, reliable(H_R, H_T) is the reliability between H_R and H_T. Fig. 1 and Fig. 2 show the agents and negotiations algorithms.

```
Agent_algorithm (hosts, comps)
{
A= Component Agent of component C_T;
B= Component Agent of component C_R;

If ( Cor(C_R,C_T)=high))
    If (Agent Negotiation (A) ==false)
    {
        Agent Negotiation (A) =true
    Result=Negotiation (A, B)
        If (Result==result1)
            Redeployment (C_R)

        If (Result==result2)
            Replication (C_R)

        If (Result==result3)
            Redeployment (C_T)

        If (Result ==result4)
            Replication (C_T)

        If (Result==result5)
            None
}}
```

```
Result=Negotiation ( Agent A, Agent B)

{
Exchange information (A,B)

If (Redeployment (C_T) ==true)
    Result 1=availability (Redeployment (C_T))

If (Replication (C_T) =true)
    Result 2=availability (Replication (C_T))

If (Redeployment (C_R)=true)
    Result 3=availability Redeployment (C_R))

If (Replication (C_R) =true)
    Result 4=availability (Replication (C_R))
Else
    Result 5=current availability
Result=max (result1, result2, result3, result4)
Return result;
}
```

Fig. 1 Agent Algorithm   **Fig. 2** Agent Negotiation

We have made some improvement on the **DeSi** simulator [6] in order to simulate interactions between any two components on different hosts. We generate a deployment architecture that consists of 10 components, 3 hosts with their software agents and with availability=.8122 distributed as follows:

Host0 = {0,4,7}, Host1 ={2,6,3,8} and Host3 = {0,5,1,9}

the input value are as in Table 1.

Table 1 Input Values

Input Parameter	Value	Input Parameter	Value
Number of Component	10	Min host reliability	0
No. of hosts	3	Max host reliability	1
Min comp memory (in KB)	2	Min comp event size (in KB)	.01
Max comp memory (in KB)	8	Max comp event size (in KB)	10
Min host memory (in KB)	15	Min host bandwidth (in KB/S)	30
Max host memory (in KB)	30	Max host bandwidth (in KB/S)	100
Min comp frequency (in events/s)	0	Level of dependency	3
Max comp frequency (in events/s)	10		

Let C^i_R where $0 \leq i \leq 9$ and C^j_T where $0 \leq j \leq 9$ be two operating components and let Rep(i) (resp. Red(i)) denote replicating C^i_R (resp. redploying) on host of C^j_T. To test the viability of the algorithms, we execute several scenarios.

The values, as generated by the simulator (of an E-deployment architecture), which effect the agents' negotiation results, are shown below: frequency values between components and their memory size in Fig. 3, dependency values in Fig. 4 and reliability values between hosts and their memory size in Table 2.

We now consider two Scenarios. In the first, C_4 makes requests frequently to C_6 (cf.Table 3). The result (cf. Fig. 5) is to replicate C_4 as there are many components dependent on it, and it provides better availability. We could not replicate C6 because there is a need for data consistency and it depends on two components in its host. In the second, C_3 makes frequents requests to C_9 (Table 4). The result (cf. Fig. 6) is that either mechanism is possible. Redeploying C_3 will improve availability because it has more interaction and both positive and negation dependency relations with components in the host of C_9.

Fig. 3 Components Frequency

Fig. 4 Component Dependency

Table 2 Host Reliablity/Memory

Host No.	0	1	2
0	1	.49	.38
1	.49	1	.94
3	.38	.94	1
Host MEM	21	27	22

Table 3 Component properties

Comp. properties	Comp (4)	Comp (6)
Comp. memory size	7,6kb	3,8
Free host size	4,5 kb	12.5 kb
Positive dependency	→0,3	→0,2,3
Data Consistency	0	1

Fig. 5 Senario 1 Results

Fig. 6 Senario 2 Results

Table 4 Scenario 1 Component properties

Component Properties	Component (3)	Component (9)
Component memory size	4.8 kb	6.8
Free Host size	4.5kb	12.5kb
Positive dependency	$\rightarrow$0,1,2	$\rightarrow$4,5,8
Negative dependency	4,6,8,9$\rightarrow$	
Data consistency	0	0

4   Previous Work and Concluding Remark

Several approaches that support the replication of components in DSs have been proposed. However, only a few address redeployment. In [4], Dock is proposed. It employs mobile agents to perform deployment tasks among hosts. It differs from our approach in that it is more concerned with the practical issues of implementing deployment rather than extracting parameters and evaluating deployment architectures. In [5], a constraint-based deployment approach is presented. It addresses the deployment of hierarchical components on heterogeneous dynamic networks. In [3], MARP employs mobile agents to coordinate the updates made to replications maintained at different servers to ensure consistency.

In this paper, we present a generic agent-based monitor approach that supports the dynamic component redeployment and replication mechanisms which were presented in Avala [6] and improved in E-Avala [2]. Some of the issues that need to be addressed include: (1) dealing with functional consistency among components, (2) expanding the solution to include additional parameters such as components structure representation.

References

1. Achmad, I., Graham, M., Santosh, K., Mark, C.: Component Replication in Distributed Systems, A Case Study Using Enterprise Java Beans. In: SRDS, pp. 89–98 (2003)
2. Al-Areqi, S., Hudaib, A., Obeid, N.: Improving Availability in Distributed Component-Based Systems via Replication. In: ACCIDS 2011, pp. 43–52 (2011)
3. Cao, J., Chan, A., Wu, J.: Achieving Replication Consistency Using Cooperating Mobile Agents. In: ICPP Workshops, pp. 453–458 (2001)
4. Hall, R., Heimbigner, D., Wolf, A.: A Cooperative Approach to Support Software Deployment Using the Software Dock. In: ICSE 1999, pp. 174–183 (1999)
5. Hoareau, D., Mahéo, Y.: Constraint-Based Deployment of Distributed Components in a Dynamic Network. In: Grass, W., Sick, B., Waldschmidt, K. (eds.) ARCS 2006. LNCS, vol. 3894, pp. 450–464. Springer, Heidelberg (2006)
6. Mikic-Rakic, M., Malek, S., Medvidovíc, N.: Improving Availability in Large, Distributed Component-Based Systems Via Redeployment. In: Dearle, A., Savani, R. (eds.) CD 2005. LNCS, vol. 3798, pp. 83–98. Springer, Heidelberg (2005)
7. Moubaiddin, A., Obeid, N.: The Role of Dialogue in Remote Diagnostics. In: 20th Int. Conf. on COMADEM (2007)

8. Moubaiddin, A., Obeid, N.: Dialogue and Argumentation in Multi-Agent Diagnosis. In: Nguyen, N.T., Katarzyniak, R. (eds.) New Chall. in Appl. Intel. Tech. SCI, vol. 134, pp. 13–22. Springer, Heidelberg (2008)
9. Moubaiddin, A., Obeid, N.: Partial Information Basis for Agent-Based Collaborative Dialogue. Applied Intelligence 30(2), 142–167 (2009)
10. Moubaiddin, A., Obeid, N.: On Formalizing Social Commitments in Dialogue and Argumentation Models Using Temporal Defeasible Logic. Knowledge and Information Systems (2012), doi:10.1007/s10115-012-0578-6
11. Obeid, N., Moubaiddin, A.: On the Role of Dialogue and Argumentation in Collaborative Problem Solving. In: ISDA, pp. 1202–1208 (2009)
12. Obeid, N., Moubaiddin, A.: Towards a Formal Model of Knowledge Sharing in Complex Systems. In: Szczerbicki, E., Nguyen, N.T. (eds.) Smart Information and Knowledge Management. SCI, vol. 260, pp. 53–82. Springer, Heidelberg (2010)
13. Osrael, J., Froihofer, L., Goeschka, K.: What service replication middleware can learn from object replication middleware. In: MW4SOC, pp. 18–23 (2006)

Prototyping and Evaluation of a Wireless Sensor Network That Aims Easy Installation

Takanobu Otsuka, Tatsunosuke Tsuboi, and Takayuki Ito

Abstract. The number of senior citizens living alone are increasing in Japan. Accordingly, the budget for social security is increasing. The percentage of burden for social security budgets reached 69.5% only for senior citizens recently, and will increase more and more. These budgets are consumed in mainly in the larger hospitals. Thus, recently "in-house" health care for senior citizens is gathering much attention in Japan. Various "home-care" products are increasingly developed and implemented to care the health of the senior citizens. However, these products are usually expensive and their self installation is very difficult. In this research, we developed a wireless sensor network system that realizes easy installation and easy operation. Our preliminary experiments demonstrate that our system can surely find some anomaly sensing information without any difficult installation procedures.

1 Introduction

The number of senior citizens living alone are increasing in Japan. Accordingly, social security budget is increasing. Even though Japans total population has become stagnant, in 2010 the percentage of senior citizens reached 23.1% and is still increasing. The percentage of burden for social security budgets reached 69.5% by senior citizens. The Japanese government realized that this is a really important budget problem and must need to be tried to be solved. The Japanese government got the medical institutions to decrease the number of beds for senior citizens because of mainly the budget, and switch the policy toward to home health care. Because of

Takanobu Otsuka
School of Engineering, University of California, Irvine, Irvine, California 92697-2625
e-mail: otsuka.takanobu@nitech.ac.jp

Tatsunosuke Tsuboi · Takayuki Ito
Nagoya Institute of Technology, Gokiso, Showa-ku, Nagoya, Aichi, Japan
e-mail: tsuboi.tatsubosuke@itolab.nitech.ac.jp,
    ito.takayuki@nitech.ac.jp

M. Ali et al. (Eds.): *Contemporary Challenges & Solutions in Applied AI*, SCI 489, pp. 27–32.
DOI: 10.1007/978-3-319-00651-2_4   © Springer International Publishing Switzerland 2013

the above situation, the importance of home health care is increasing drastically in Japan. Various home care products are developed to provide home health care for senior citizens. The following are the examples of them.

- The alert system that monitors for abnormally wanders by senior citizens
- Life safety alarm system to monitors for an old person living alone
- Emergency call and reporting systems for unusual status
- Emergency telephone calling system for unusual status

However, these products are usually expensive and their installation is difficult. Our research aims to develop a system for easy operation and easy installation. Because we conducted a cooperative research with the Niihama medical Coop as a preliminary test in Japan. In this research, we could identify the problem in which system developer needs to know in the real world. We conducted a research with the cooperation of the Niihama Medical Co-op in Japan. Niihama Medial Co-op was consistently operates a large clinic, a group home, a day care, and a day service. It is a few medical institutions in Japan which gives home health care and medical care throughout. Generally medial institutions are divided medical services and home health care services. Therefore the burden on user is big the Niihama Medical Co-op. was integrating medical and home health care services. The Japanese government got the medical institutions to decrease the number of beds for senior citizens because of mainly the budget, and switch the policy toward to home health care. In this research we aim to realize an easy-install and easy-operation sensor modules for both of the care-provider sides and senior people.

The rest of our paper is organized as follows. Section 2 introduces previous studies and the position of our research. Section 3 presents our prototype system and the configuration of our test demonstration. Finally, Section 4 summarizes our paper and provides future work.

2   Related Works

2.1   General Product Features

The system [1] can simply notify 2 times per day for the senior citizen when they drink a cup of tea. The systems [2]can analyze the behavior patterns using the motion sensors and send some emergency reports.[6][7] The systems [4]have been developed with RFIDs and sensors for analyzing the behavior patterns in order to send the notifications of abnormal events as a whistle-blower system. The systems[3]have the wearable sensors that can monitor temperature by the thermometers. The system [5]can detect abnormal behavior by using a video camera. The following are the problems on the above related works.

- Generally they are expensive products for installation.
- They require the separated communication cables.

- Video cameras will invate privacy too much.
- Werable sensors increase the mental burden of users
- They require huge computational power to process the videos
- Their main purposes are to altert rathern to wantch/see.
- Their visibility of the activity logs is not sophisticated.

3 Development and Test Demonstration of WSNs

3.1 Feature of Zigbee

We selected the most suitable transmission standards because it should be wireless.

The most unique point of Zigbee is that itfs ultra-low power consumption and it can built easily for a mesh type network[8][9]. Specifically, Zigbee-based modules work 6 monthes with only the battery type CR2032. When a communication command is taken out from the unit A to the unit B, it can select a single route from A to B. You can select any route for data coomunication in the mesh network of Zigbee.

The second unique point is the mesh network of Zigbee. Installed Zigbee has a repeater function, so that you can operate as a repeater by setting Zigbee in the middle point in the case of short of communication distance. It is the optimal as a sensor network because you can use these features without special installation. Moreover some Zigbee have a program area alone, you can mount the sensor which easy calculation is attached although rich calculation canft be performed.

3.2 The System Outline

We have developed a sensor network that can be installed easily by anyone. While users can begin to operate by simply putting unit, the sensing result can be viewed on the webpage. This section describes the developed system. An overview of the system is shown in Figure1.

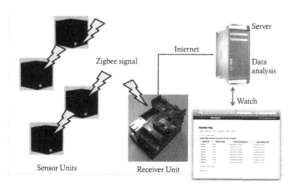

Fig. 1 System outline

Sensed data by the sensor units transmits to the server. The server can display the sensed data in a web page almost realtimely. We developed the sensor unit is running by an internal battery. This is because any power supply cable is not required.

3.3 Experimental Setting

By installing the sensor units for two of our rooms, we have collected the sensed data for performing anomaly detection by sensing the actual data. The sensor unit has been placed in accordance with the flow line of the students.

While Figure 2 shows the room-A layout, Figure 3 shows the room-B layout.

Room-A is divided into two main areas, illustrated the flow line to the path toward the entrance of their desk. Sensors are arranged at right angles to the flow line thereof, because it is about 5m, the entire laboratory sensingis possible. For simplicity, room A is similar to that of a relatively businesses and households.

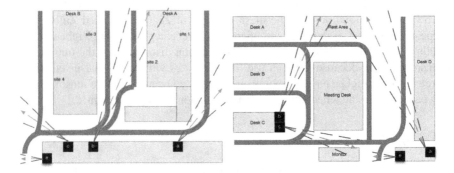

Fig. 2 Room-A layout diagram of the sensor unit

Fig. 3 Room-B layout diagram of the sensor unit

Room-B is a meeting room space in which, there is a desk that is arranged at the center, and the other tables surround it. We assume room-B is a conference room or a free space.

We collected the following data:

- Data collection periodF2012/07/01-2012/08/22
- Number of data collection of room AF242,053
- Number of data collection of room BF177,623

Our outlier detection is performed by clustering the collected data. We divided into four clusters by the k-means method. While the vertical axis as the time, the horizontal axis are date. A data that deviates from clusters can be seen as a anomaly.

3.4 Experimental Result

Experimantal results for room-A and room-B are shown in Figure 4 and Figure 5, respectively. Because there are a lot of people in the room-A compared to room B, while the result on room-A looks more continuous, the result of room-B is more intermittent but looks continuous as well. From both figures, many sequential reactions of sensor e, sensor b and sensor c can be seen in the results because they were installed in a doorway where many people were passing.

In the Figure 4 and Figure 5, some anomaly data can be seen. These anomaly data are actually reactions caused by a heat source by the sunlight. However, what we can confirm is our sensor system can show some anomaly data in the graphs. In the real usecase, analyist can interpret these anomaly data based on the domain.

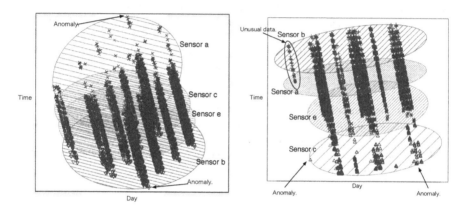

Fig. 4 Clustering result of room A   **Fig. 5** Clustering result of room B

4 Summary

Recentry, a lot of services that can monitor senior people or detect intruders have been focused very much. But the usual systems are expensive and we need expert knowledge to install them. In this study, we have developed a sensor network that is low cost and easy to installation. In addition, the system also performs the anomaly detection method by clustering. We conducted the experiments with 2 real typical rooms. Then, we showed that our system can classify correctly the normal sensed data and abnormal sensed data. The abnormal sensed data, in this experiment, was a kind of the noise[10]. But it could be an intruder or some other happening. As future works, realization of anomaly detection using bayesian network, it is possible to aims easy installation.

Acknowledgements. This work is partially supported by the Funding Program for Next Generation World-Leading Researchers (NEXT Program) of the Japan Cabinet Office.

References

1. Zoujirushi Corp., Mimamori Hot Line, `http://www.mimamori.net/`
2. Aoki, S., Onishi, M., Kojima, A., Sugahara, Y., Fukunaga, K.: Recongnition of a Solitude Senior's Behavioral Pattern Using Infrared Detector, The institute of Electronics, Information and Communication Engineers. Technical report No. 2001-50 (2002)
3. Tanaka, H., Nakauchi, Y.: Senior Citizen Monitoring System by Using Ubiquitus Sensors. Society of Mechanical Engineers, Journal No. 75–760 (2009)
4. Furuya, M., Murakami, H., Miyamoto, W.: Detecting of an illness date based on usual activities of the single elderly person by using a few sensors, The institute of Electronics, Information and Communication Engineers, Technical report No. 2002-125 (2003)
5. Seki, H., Hori, Y.: Detection of Abnomal Action Using Image Sequence for Monitoring System of Aged People, The institute of Electronics, Information and Communication Engineers, Jouanal No. 122–2 (2000)
6. Chandola, V., Banerjee, A., Kumar, V.: Anomaly Detection: A Survey. Technical Report, Department of Computer Science and Engineering University of Minnesota, TR-07-017 (2007)
7. Rajasegarar, S., Leckie, C., Palaniswami, M.: Distributed Anomaly Detection in Wireless Sensor Networks. In: 10th IEEE Singapore International Conference on Communication Systems, ICCS (2006)
8. Lennvall, T.: A comparison of WirelessHART and ZigBee for industrial applications. In: Factory Communication Systems, WFCS 2008, May 21-23 (2008)
9. Varchola, M., Drutarovsky, M.: Zigbee Based Home Automation Wireless Sensor Network. Acta Electrotechnica et Informatica 7(4) (2007)
10. Rousseeuw, P.J., Leroy, A.M.: Robust regression and outlier detection. John Wiley and Sons, Inc., New York (1987)

Part III
Evolutionary Algorithms

Winner Determination in Combinatorial Reverse Auctions

Shubhashis Kumar Shil, Malek Mouhoub, and Samira Sadaoui

Abstract. Since commercially efficient, combinatorial auctions are getting more interest than traditional auctions. However, winner determination problem is still one of the main challenges of combinatorial auctions. In this paper, we propose a new method based on genetic algorithms to address two important issues in the context of combinatorial reverse auctions: determining the winner(s) in a reasonable processing time and reducing the procurement cost. Indeed, not much work has been done using genetic algorithms to determine the winner(s) specifically for combinatorial reverse auctions. To evaluate the performance of our method, we conducted several experiments comparing our proposed method with another method related to determining winner(s) in combinatorial reverse auctions. The experiment results clearly demonstrate the superiority of our method in terms of processing time and procurement cost.

1 Introduction

An auction is a market scenario in which bidders compete for item(s). In traditional auctions, an individual item is auctioned separately, which leads to an inefficient allocation and processing time [7, 10]. Combinatorial auctions have been proposed to improve the efficiency of bid allocation by allowing bidders to bid on multiple items [7, 10]. These auctions provide a combinatorial allocation that minimizes the procurement cost and running time [5, 7, 10]. They have been used in various real-world situations [1] such as resource allocation with real-time constraints [10], sensor management [9, 11], supply chain management [12] and computer grids [2]. A combinatorial auction problem is actually a winner determination problem [4]. Winner determination is still one of the main challenges of combinatorial auctions [10]. Indeed, determining the winner(s) in combinatorial

Shubhashis Kumar Shil · Malek Mouhoub · Samira Sadaoui
Department of Computer Science, University of Regina, Regina, SK, Canada
e-mail: {shil200s,mouhoubm,sadaouis}@uregina.ca

M. Ali et al. (Eds.): *Contemporary Challenges & Solutions in Applied AI*, SCI 489, pp. 35–40.
DOI: 10.1007/978-3-319-00651-2_5   © Springer International Publishing Switzerland 2013

auctions is a complex problem and it has been shown to be NP-complete [5, 7]. However, applying combinatorial auctions to procurement scenario [13] is cost–saving [7]. Many algorithms have been developed to solve combinatorial auction problems, e.g. Hsieh and Tsai presented a Langrangian heuristic method [7], and Sitarz introduced Ant algorithms and simulated annealing [7]. Furthermore, many research works have been carried out to figure out the efficient way to solve winner determination in combinatorial auctions. Most of the proposed algorithms restrict the bundles on which bids can be submitted in order to solve the problem optimally but these restrictions introduce economic inefficiencies [7]. Some algorithms find optimal solutions but are very slow; others avoid restrictions but allow bidding on a small number of items [7]. We are interested in combinatorial reverse auctions in which we consider the procurement of a single unit of multi-items. In our auction, there is one buyer and several sellers who compete according to the buyer's requirements. First the buyer announces his demand (multiple items) in the auction system. Then the interested sellers register for that auction and bid on a combination of items. Genetic Algorithms (GAs) are successful to solve many combinatorial optimization problems [7]. GAs are powerful search techniques consisting of selection, crossover and mutation methods [3]. Sometimes simple crossover and mutation operators produce inappropriate chromosomes. To avoid this problem, special or modified crossover and mutation operators are defined [1]. GAs can terminate anytime as required and the current best chromosome can be the best solution. Nevertheless, not much work has been done by using GAs to solve winner determination problem in the context of combinatorial reverse auctions. To the best of our knowledge, only one research paper [7] employed GAs to tackle this problem. However, the method proposed in [7] needs comparatively many generations and a considerable amount of time to produce good solutions. In this paper, our research goal is twofold: (1) solve the winner determination problem in combinatorial reverse auctions in a reasonable processing time, and (2) reduce the procurement cost with fewer generations. For this purpose, we define a new GA-based method that uses two repairing techniques to repair infeasible chromosomes as well as a modified two-point crossover operator that is capable of distributing the solutions and preventing a premature convergence. Furthermore, we conduct several experiments by comparing our proposed method with the one defined in [7]. The experimental results clearly demonstrate the superiority of our method in terms of processing time and procurement cost.

2   Proposed GA-Based Method

In Fig. 1, we define our GA-based method that we name GACRA (Genetic Algorithms for Combinatorial Reverse Auctions). Assume there are m items and n sellers. So the number of bid items combination is 2^m-1 and we use m×n bits to represent each chromosome. In case of m=2 and n=3, a chromosome represented by 100100 means that seller 1 bids for only item A (first two bits 10), seller 2 bids for only item B (next two bits 01) and seller 3 bids for no items (last two bits 00). To generate bid prices, we consider random values between 200 and 500 for each item for each seller. In Step 2, the initial chromosomes are generated randomly. To avoid redundancy, the RemoveRedundancy function ensures exactly one selection of a particular item from all sellers in each chromosome. To avoid emptiness, the RemoveEmptiness function guarantees at least one selection of every item in each chromosome. So, we repair infeasible chromosomes using these two functions. Our RemoveRedundancy function works with the following steps.

1. For each chromosome, selects bits for each seller.
2. Tests bits for one seller to verify if item(s) are selected and stores this information.
3. Continue testing bits of next sellers; if item(s) are already selected by the previous seller then converts the current bit value to 0.

Our RemoveEmptiness function works with the following steps.

1. For each chromosome, selects bits for each seller.
2. Tests bits for all sellers and stores the information of the non- selected item(s).
3. Continue converting bit value to 1 until all item(s) are selected.

Input: number of bid items: m, number of sellers: n, number of generations: δ, crossover rate: α, mutation rate: β

Terminology: number of chromosomes in a population: n × m, number of combinations of bid items: 2^m-1

Features: RemoveRedundancy method, RemoveEmptiness method, modified two-point crossover

Consideration: minimize bid price in optimal running time

Output: winner(s)

1. generate bid prices for each combination of bid items for each seller
2. generate initial chromosomes
3. check feasibility and remove redundancy and emptiness, if necessary

Loop until number of generations achieved

{

4. compute fitness values of chromosomes
5. select chromosomes with gambling-wheel disk selection method
6. compute fitness values of chromosomes
7. generate child chromosomes from parents with modified two-point crossover considering crossover rate, α
8. mutate chromosomes considering mutation rate, β
9. check feasibility and remove redundancy and emptiness, if necessary
10. compute fitness values of chromosomes
11. determine better chromosomes from both initial and new chromosomes of each generation

}

12. return winner(s) with minimum bid price in optimal running time

---

Fig. 1 Pseudo code of GACRA

In step 4, fitness value of chromosome is generated. We propose the following fitness function to calculate the fitness value of every chromosome. Since the motivation of this research work is to minimize the procurement cost for the buyer, our fitness function for Chromosome X_i is defined as follows.

$$F(X_i) = \frac{1}{\sum_{s=1}^{n}\sum_{C=1}^{2^m-1} b_s(C) \times x_s(C)} \qquad / x_s(C) \in \{0,1\} \qquad (1)$$

where $b_s(C)$ represents a bid for the item combination C submitted from the sth seller; $x_s(C)$ is 1 when the item combination C is selected for the sth seller and 0 otherwise.

We use gambling-wheel disk selection method [4] to select chromosomes for cross-over operation. In step 7, the crossover operation is performed. A child chromosome takes two portions from one parent and one portion from another parent in two-point crossover. In our modified two-point crossover operation, the basic idea is same but the direction of taking portions from parents is different. The first child takes portions in forward direction but the second child takes it in reverse order. It creates positive effect to increase diversity in the solution spaces to allow all bidders to get more chances to be selected. In some cases, after removing redundancy and emptiness, some bidders get deprived but this modification of crossover gives them a chance again. This will prevent the procedure to converge prematurely. Then, the procedure will move to mutation operation as indicated in our algorithm. In step 9, RemoveRedundancy and RemoveEmptiness functions remove redundancy and emptiness respectively. In step 11, the procedure selects the better chromosomes among the initial and new chromosomes of the generation based on fitness values. Since genetic algorithm is called anytime algorithm, our procedure can be stopped anytime and it produces the best solution. The entire process is repeated until the termination condition is fulfilled, which is here the number of generations. In step 12, the procedure returns the winner(s). The solution is not improving in all generations but in our procedure we always preserve the current winner. So there is no chance to produce worse solution than the previous generation.

3 Experiment

We have conducted several experiments to determine the winner in combinatorial reverse auctions by using our method GACRA. We also compare GACRA with the technique presented in [7] that we call CRA (Combinatorial Reverse Auctions). We have implemented GACRA as well as CRA as described in [7] in Java. These two methods are both executed on an AMD Athlon (tm) 64 X2 Dual Core Processor 4400+ with 3.43 GB of RAM and 2.30 GHz of processor speed. We have used the following parameters and settings for the experiments: Chromosome Encoding: Binary String; Selection: Gambling-Wheel Disk; Crossover: Modified

Two-point; Crossover Rate: 0.6; Mutation Rate: 0.01 and Termination Condition: Generation Number.

In the first experiment, we measure the required time of our proposed method and compare it with CRA. In Fig. 2, we show the required time (in milliseconds) versus the number of generations for both GACRA and CRA. This is the average required time of 20 runs. From the comparison we can see that our proposed method needs less processing time. This happens because of two reasons: (1) we represent the chromosome with less number of bits, and (2) we keep the calculation of fitness value simple.

We have also done some comparative experiments on the procurement cost and report the results in Fig. 2. Since our procedure always maintains feasible solutions and never accepts redundant bid item, it is able to produce good solutions from the very first generations. Moreover, it keeps producing better solutions in consecutive generations.

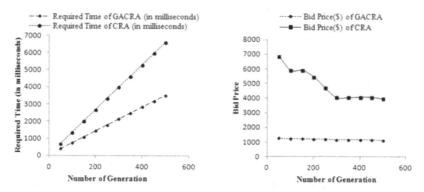

Fig. 2 Left figure - required time (in milliseconds) vs number of generation; right figure – bid price vs number of generation

In addition, we assess the processing time for GACRA by varying the number of sellers and keeping the number of items fixed to 2, 4, 6 and 8. Finally we assess the processing time for GACRA by varying the number of items and keeping the number of sellers fixed to 20, 40, 60 and 80. As the number of bits required to represent the chromosomes of our algorithm directly depends on both the number of items and the number of sellers, so with increasing any of these, the required running time also increases. Due to using less number of bits to represent chromosome, required running time of GACRA is less than that of CRA. In CRA fitness function tries to mitigate the effect of allowing redundancy but in GACRA fitness function is simple and has no additional task. In CRA repair function only removes emptiness but in GACRA RemoveRedundancy, RemoveEmptiness functions and Modified two-point crossover operator remove redundancy, emptiness and early convergence respectively. Due to not allowing redundancy, GACRA is able to minimize procurement costs from very early generations.

4   Conclusion and Future Work

Motivated to save significant time and reduce the procurement cost, this paper solves the problem of winner determination in combinatorial reverse auctions. With the help of the two repairing method, the modified two-point crossover and a careful selection of operator of GAs, it is notable that our method can produce optimal solutions in a reduced processing time. As it is obvious that parallel GAs are capable of providing the solution more efficiently [6, 8], the future target of this research is to determine the winner by using the most efficient methods of parallel GAs. Another future direction is to determine the winner(s) according to multiple attributes of items.

References

1. Easwaran, A.M., Pitt, J.: An Agent Service Brokering Algorithm for Winner Determination in Combinatorial Auctions. In: ECAI, pp. 286–290 (2000)
2. Das, A., Grosu, D.: A Combinatorial Auction-Based Protocols for Resource Allocation in Grids. In: 19th IEEE International Parallel and Distributed Processing Symposium (2005)
3. Goldberg, D.E., Deb, K.: A Comparative Analysis of Selection Schemes Used in Genetic Algorithms, pp. 69–93 (1991); edited by G.J.E. Rawlins
4. Gong, J., Qi, J., Xiong, G., Chen, H., Huang, W.: A GA Based Combinatorial Auction Algorithm for Multi-Robot Cooperative Hunting. In: International Conference on Computational Intelligence and Security, pp. 137–141 (2007)
5. Zhang, L.: The Winner Determination Approach of Combinatorial Auctions based on Double Layer Orthogonal Multi-Agent Genetic Algorithm. In: 2nd IEEE Conference on Industrial Electronics and Applications, pp. 2382–2386 (2007)
6. Nowostawski, M., Poli, R.: Parallel Genetic Algorithm Taxonomy. In: Third International Conference on Knowledge-Based Intelligent Information Engineering Systems, pp. 88–92 (1999)
7. Patodi, P., Ray, A.K., Jenamani, M.: GA Based Winner Determination in Combinatorial Reverse Auction. In: Second International Conference on Emerging Applications of Information Technology, pp. 361–364 (2011)
8. Abbasian, R., Mouhoub, M.: An Efficient Hierarchical Parallel Genetic Algorithm for Graph Coloring Problem. In: 13th Annual Genetic and Evolutionary Computation Conference, pp. 521–528. ACM, Dublin (2011)
9. Mullen, T., Avasarala, V., Hall, D.L.: Customer-Driven Sensor Management. IEEE Intelligent Systems 21(2), 41–49 (2006)
10. Avasarala, V., Polavarapu, H., Mullen, T.: An Approximate Algorithm for Resource Allocation using Combinatorial Auctions. In: International Conference on Intelligent Agent Technology, pp. 571–578 (2006)
11. Avasarala, V., Mullen, T., Hall, D.L., Garga, A.: MASM: Market Architecture or Sensor Management in Distributed Sensor Networks. In: SPIE Defense and Security Symposium, Orlando FL, pp. 5813–5830 (2005)
12. Walsh, W.E., Wellman, M., Ygge, F.: Combinatorial Auctions for Supply Chain Formation. In: ACM Conf. on Electronic Commerce, pp. 260–269 (2000)
13. Narahari, Y., Dayama, P.: Combinatorial Auctions for Electronic Business. Sadhana 30(Pt. 2 & 3), 179–211 (2005)

Virus Transmission Genetic Algorithm

Weixin Ling and Walter D. Potter

Abstract. In this paper we propose a novel Virus Transmission Genetic Algorithm, which is inspired by the evolution of immune defense and the infection transmission model. Containing one virus population and one host population, the VTGA simulates biological infections by using new operators such as virus infection and virus spread. To study the effectiveness, we apply the algorithm to several function optimization problems, several travelling salesman problems and a forest planning problem. Results of the experiments show that the VTGA performs well at searching for optimal solutions and preserving diversity of population.

Keywords: Virus Transmission Genetic Algorithm, Function Optimization, Travelling Salesman Problem, Forest Planning Problem.

1   Introduction

Premature convergence is a well-known problem of the Genetic Algorithm [1] [2] [3]. One solution is storing patterns during a search. Hence, diversity won't be eliminated when the algorithm converges. The Virus-Evolutionary Genetic Algorithm [4] [5] is an algorithm implementing this method. The VEGA has a virus population and a host population. The virus population is for saving effective patterns. When a new generation of hosts is created, these effective patterns offer information to make the new hosts better than their parents. Reasonable as it sounds, there are problems with the VEGA. The first one is that every virus represents a continuous chromosome region, not a complete solution. If the effective patterns don't exist as continuous regions, then the viruses can't offer much help. The second problem is how to determine the fitness of the viruses. Since the virus population has a limited size. Hence, only the most effective patterns will be stored and we need to evaluate patterns to decide which ones will stay. In the

Weixin Ling · Walter D. Potter
Institute for Artificial Intelligence, University of Georgia, United States

M. Ali et al. (Eds.): *Contemporary Challenges & Solutions in Applied AI*, SCI 489, pp. 41–46.
DOI: 10.1007/978-3-319-00651-2_6   © Springer International Publishing Switzerland 2013

VEGA, the fitness value of a virus measures how much the virus improves the hosts infected by it. So, the evaluation of a virus involves many evaluations of the hosts related to the virus, which could be computationally expensive for some complex problems.

Consequently, considering the reasonableness of storing patterns and the problems of the VEGA, we propose a new algorithm called the Virus Transmission Genetic Algorithm. The VTGA uses two populations as well, one virus population and one host population. However, they work differently. The virus population is used for search and the host population is used for storing good schemes. In the VTGA, we simulate the evolution of viruses inside their hosts. Therefore, we have crossover and mutation working similarly to that of the simple GA. To mimic the behaviors of viruses, we implement infection, spread and recovery operators following the Susceptible-Infectious-Recovered (SIR) model in epidemiology [6].

In the VTGA, we involve discussion about the biological immune system. However, the VTGA is an algorithm very different from the well-known Artificial Immune System (AIS), which abstracts the structure and function of the natural immune system through pattern recognition techniques [7].

In the remainder of the paper, we provide a description of the VTGA. Then we examine the efficiency of our algorithm by using it to solve some function optimization problems, some travelling salesman problems and a 73-stand forest planning problem. Finally, we discuss the features of the algorithm and future work.

2   Virus Transmission Genetic Algorithm

To model the evolution of immune defenses against infectious diseases, one host population and one virus population are defined in the VTGA. Although named differently, host individuals and virus individuals both represent a complete solution. The VTGA initializes hosts and viruses by assigning random solutions to them, which are evaluated by the same objective function. The host population is mainly used for recording good solutions found during computation. No operation is defined for the host population and there is no interaction among hosts. Virus individuals, on the other hand, perform many operations, the outline of which is this: There are attacks from virus individuals to hosts. Weak hosts are killed by viruses and replaced by stronger hosts. As a result the overall fitness of the host population is gradually improved. To simulate this process, we need to implement virus infection, virus evolution (crossover and mutation) and virus spread.

For infection, a virus selects a weak host adopts information from the host. We use a tournament selection to randomly pick up a few candidate hosts. The candidate host with the worst fitness value will be selected as the target, following the fact that people with weak immune systems are likely to get sick. When solving an optimization problem, we find that it's usually harder to improve better solutions than worse ones. Hence, with the "selecting the worst scheme", we improve poor solutions first. After selection, the virus will adapt itself to the host. In biology, adaptation is a complex process. A virus replicates and mutates to generate lots of strains to beat the host's immune system. But in this study, we simplify this process by copying several genes from the host, which could be seen as a process of bringing effective patterns from the host to the virus. The parameter INFECTION_RATE is used to determine the percentage of genes being copied.

After infection, a virus starts developing itself by crossover and mutation inside its host. Only viruses infecting the same host could be matched together for crossover, so the viruses infecting the same host form a dynamic subpopulation, which is isolated from other subpopulations of other hosts and whose size is changing because some of the viruses will leave and infect another host from time to time. For this reason, although the size of the whole virus population remains the same, the virus subpopulation size inside a host is changing all the time. In this case, it is difficult to use traditional crossover operators. Because they usually require parent selection and replacement and implementing selection and replacement on dynamic virus subpopulations in different hosts is complicated.

Considering that, we replace the traditional selection process with enumerating every pair of two virus individuals inside the same host and exchanging information between them. The crossover process is one-way. The better virus copies some random genes to replace the corresponding genes of the worse virus but not the other way around. How many positions are copied is controlled by the parameter CROSSOVER_RATE. After the weaker virus is rewritten, the newly generated virus will not be evaluated and its fitness value remains the same. The crossover is finished when every pair of viruses of the same host is processed.

We gain two advantages from this method. Inside the same host, the best virus has influence on every other virus. The second best virus has influence on every other virus except the best one and so on. On the other hand, the worst virus will be affected by all other viruses. And the second worst virus will be affected by all other viruses except the worst. That means better solutions could have an influence on more solutions and worse ones cause impacts on fewer solutions.

After crossover, the VTGA mutates viruses by reassigning random values to genes with a low probability, which is controlled by a parameter named MUTATION_RATE. When mutation is finished, all viruses will be evaluated.

In an appropriate situation, a virus escapes from its current host and selects another one. Virus spread as an operator simulates this process of escaping. After crossover and mutation, the VTGA goes through all viruses to check their fitness values. If a virus is good enough to defeat its host by having a better fitness, then there is a chance that the host's solution will be replaced completely by the solution of the virus. Whether a host will be replaced or not is controlled by a parameter called SPREAD_RATE. In another case when a virus is not better than its host, the VTGA allows the virus to escape with a probability controlled by a parameter called the RECOVERY_RATE. Once a virus escapes from its host, we would consider the host's immune system to have recovered from the infection of this virus so that the virus won't perform crossover or mutation until it infects next host.

In the VTGA, the infection operator randomly selects a few candidates from the host population, and a virus will infect the weakest one of the candidates. So the worse a host is the more opportunities it has to get infected. But there is one problem. Suppose the VTGA selects N candidates from the host population in every infection. If N is smaller than the host size, some hosts will never be selected because they are the best in the population. So the VTGA spends all its effort on improving relatively bad solutions. This mechanism gradually raises the lower bound of all host fitnesses. However, in optimization problems, this is not efficient because the goal is to find the best solution. So we use a method called fitness masking on the best host. At the end of every generation, the VTGA finds the best individual in the population. After the VTGA records the best individual, the fitness masking operator randomly selects another fitness value from the host population to replace the fitness value of the best individual in the population.

3   Experiment Results

Function Optimization. At the first stage of experiments we have six quality tests using several famous test functions [8]. The six functions are: De Jong's function, Axis parallel hyper-ellipsoid function, Rotated hyper-ellipsoid function, Rosenbrock's valley, Rastrigin's function and Schwefel's function.

Travelling Salesman Problems. The travelling salesman problem (TSP) is a well-known NP-hard problem in combinatorial optimization. In this study, we test the VTGA with several TSP problems (Bays29, Swiss42, Berlin52, Eil76 and Ch150) obtained from the TSPLIB[1]. We describe a solution of the TSP problems as a permutation of integers. And we use the order based crossover operator proposed by Syswerda [9] [10] and the exchange mutation operator.

The Forest Planning Problem. The goal of solving a forest planning problem [11] is to find the best valid harvest schedule maximizing the even-flow of harvest volume. We use the VTGA to solve the 73-stand Daniel Pickett Forest[2] [11].

The forest planning problem is a constrained optimization problem. A valid solution of the problem has to obey all rules listed in the problem. To acquire valid solutions using a search algorithm, we repair invalid solutions after mutation so that they become violation-free before they are evaluated.

Results. We run 50 trials for every function optimization problem. From the results (Table 1), we learn that the VTGA is able to obtain good results in most tests.

We use different configurations when solving the TSP problems (Table2). From the results (Table 2), we believe the VTGA is an effective algorithm for solving the TSP problems generally.

For comparison, we implement one generational GA (Table 3) with the same repair function. We find that after 1 million evaluations, it is difficult for the GA to improve the solutions it finds. The best solution found by the GA remains the same for many generations before any improvement. But the VTGA is able to improve its best solution more or less after every 1,000,000 evaluations. It outperforms the GA on the average fitness and the best fitness. And the VTGA achieves 69.2301% accuracy of finding the global optimum in 30 trials.

---

[1] http://www.iwr.uni-heidelberg.de/groups/comopt/
software/TSPLIB95/

[2] Data could be found at:
http://www.warnell.forestry.uga.edu/Warnell/Bettinger/
planning/index.htm

Table 1 Results for the Function Optimization Problems[3]

Function	N	Global Best	20000 Evaluations		100000 Evaluations	
			Best	Average	Best	Average
De Jong's function	5	0	3.52E-17	1.6E-05	0	2.8E-22
Axis parallel hyper-ellipsoid	5	0	1.90E-18	9.8E-06	0	1.2E-22
Rotated hyper-ellipsoid	5	0	6.45E-14	0.00167	0	1.1E-20
Rosenbrock's valley	5	0	0.13	2.66363	0.12	2.54
Rastrigin's function	5	0	0	3.21E-07	0	0
Schwefel's function	5	-2094.91	-1787.40	-1451.82	-2094.71	-2010.84

Table 2 Results for the Travelling Salesman Problems

TSP Instance	HS	VS	CR	MR	IR	SR	RR	Global Optimum	Best	Average Fitness	STD	Avg. Number of Evaluations (Million)
Bays29	200	1000	0.3	0.01	0.7	1.0	0.8	2020	2020	2020.2	1.08	0.34
Swiss42	200	1000	0.3	0.01	0.7	1.0	0.8	1273	1273	1274.7	5.92	2.5
Berlin52	200	1000	0.3	0.01	0.7	1.0	0.8	7542	7542	7570.3	56.89	7.1
Eil76	200	1000	0.3	0.01	0.9	1.0	0.5	538	538	548.9	5.62	10.8
Ch150	200	1000	0.1	0.01	1.0	1.0	0.8	6528	7107	7803.7	367.56	14.7

Table 3 Comparison between the VTGA[4] and the GA[5] on Forest Planning

Algorithm	Average Fitness	The Best Fitness	The Number of Trials Finding the Global Optimum
VTGA	**5794786.28011043**	**5500330.279304971**	**21**
GA	8107453.58754281	5502420.092225004	0

4   Conclusion and Future Work

Favoring Bad Solutions. Traditional GA favors good solutions by giving them more opportunities for recombination. Considering that improving good solutions is harder and harder while their fitness values increase, we implement a mechanism draws more attention to improving bad solutions.

---

[3] Host Size = 10, Virus Size = 1000, Crossover Rate = 0.7, Mutation Rate = 0.01, Infection Rate = 0.2, Spread Rate = 0.2 and Recovery Rate = 0.8.

[4] Host Size = 10, Virus Size = 1000, Crossover Rate = 0.3, Mutation Rate = 0.01, Infection Rate = 0.9, Spread Rate = 0.8 and Recovery Rate = 0.8.

[5] Population = 1000, Selection = Tournament, Crossover = 2-point, Crossover Probability = 0.8, Mutation = Random Resetting, Mutation Probability = 0.4, Mutation Rate = 0.01, Elitism = Yes and Repair = Yes.

Isolated Evolution. In the VTGA, viruses are isolated within different hosts. Evolution, therefore, involves only viruses in the same host. Through experiments, we find that this mechanism helps with maintaining diversity in the population.

Different Implementation of Operators. We implement a one-way crossover sending genetic information only from strong viruses to weak viruses. And the way this operator delivers information is similar as the way uniform crossover performs in a traditional GA. We think there are other ways to implement operators.

The Role of the Algorithm. Because the VTGA uses the host population to record good solutions of different areas of the search space, it preserves diversity well. Besides, the GA can only return one solution after it converges. So people generally don't have many choices. However, the VTGA returns many diverse solutions with relative good fitness values when it finishes one computation, which helps us understand the search space of a problem better by giving us representative solutions from different areas and gives us flexibility to adjust our decisions.

References

[1] Holland, J.H.: Adaptation in natural and artificial systems. University of Michigan Press, Ann Arbor (1975)

[2] Davis, L.D.: Handbook of Genetic Algorithms. Van Nostrand Reinhold, New York (1991)

[3] Goldberg, D.E.: Genetic Algorithms in Search, Optimization, and Machine Learning. Addison-Wesley, Reading (1989)

[4] Kubota, N., Fukuda, T., Shimojima, K.: Virus-evolutionary gentic algorithm for a self-organizaing manufacturing system. Computers & Industrial Engineering 30(4), 1015–1026 (1996)

[5] Kubota, N., Shimojima, K., Fukuda, T.: The role of virus infection in virus-evolutionary genetic algorithm. In: Proceedings of IEEE International Conference on Evolutionary Computation, pp. 182–187 (1996)

[6] Anderson, R.M., May, R.M.: Population biology of infectious diseases: Part 1. Nature 280, 361–367 (1997)

[7] de Castro, L.N., Timmis, J.: Artificial Immune Systems: A New Computational Intelligence Approach. Springer, London (1996)

[8] Molga, M., Smutnicki, C.: Test functions for optimization needs (2005), http://www.zsd.ict.pwr.wroc.pl/files/docs/functions.pdf

[9] Larrañaga, P., Kuijpers, C.M., Murga, R.H., Inza, I., Dizdarevic, S.: Genetic Algorithms for the Travelling Salesman Problem: A Review of Representations and Operators. Artificial Intelligence Review 13(2), 129–170 (1999)

[10] Syswerda, G.: Schedulel Optimization Using Genetic Algorithms. In: Handbook of Genetic Algorithms, pp. 332–349. Van Nostrand Reinhold, New York (1991)

[11] Bettinger, P., Zhu, J.: A new heuristic for solving spatially constrained forest planning problems based on mitigation of infeasibilities radiating outward from a forced choice. Silva Fennica 40(2), 315–333 (2006)

Automated Phenotype-Genotype Table Understanding

Shifta Ansari, Robert E. Mercer, and Peter Rogan

Scholarly writing in the broad area of experimental biomedicine is a genre that has a rhetorical style that exhibits some easily identifiable stylistic features: division of the paper into well-defined sections (Introduction, Methods, Results, Discussion), and the use of tables and figures to organize and express important results. Tables and figures have stylistic features, as well: titles, captions, content.

In addition to these common stylistic features, the community-accepted rhetorical style for authors of scientific papers is to publish their experimental findings in a tabular form, because the quantity of experimental data is large and the tabular arrangement allows for a concise presentation of the relationships among the data and for a rapid understanding of the results. Thus, tables are one of the most important sources of information.

The rapid advancement of knowledge in the biomedical field [1] has led to significant efforts to extract information from papers and interpret it automatically. Our contribution to this effort is a general approach not only to access and extract information from tables, but also *to understand* the information contained in tables by semantically grounding it with the appropriate concepts in an ontology, and to make it available for further use.

We report here our phenotype-genotype table understanding undertaking. To summarize: We have curated papers from the domain of genetics that discuss phenotype, genotype, mutation, and gene, and their relationships, syndrome (constellation of phenotypes), and disease, to design an ontology that captures the concepts needed to understand these tables and to engineer a tool which populates this ontology with data reported in these tables.

Shifta Ansari · Robert E. Mercer
Department of Computer Science, The University of Western Ontario, London, ON, Canada
e-mail: sansar6@uwo.ca, mercer@csd.uwo.ca

Peter Rogan
Department of Biochemistry, The University of Western Ontario, London, ON, Canada
e-mail: progan@uwo.ca

M. Ali et al. (Eds.): *Contemporary Challenges & Solutions in Applied AI*, SCI 489, pp. 47–52.
DOI: 10.1007/978-3-319-00651-2_7   © Springer International Publishing Switzerland 2013

Table 1: Clinical features and size of deletion of the 12 patients with 13q monosomy.

patients	1	2	3	4	5	6	7	8	9
deleted segment	13q13.3-13qter	13q21.1-13q31.1	13q21.32-13qter	13q31.1-13q33.3	13q31.1-13qter	13q31.1-13qter	13q31.3-13q31.1	13q31.3-13q34	13q32.1-13qter
size of deletion	70 mb	30 mb	47 mb	28 mb	34 mb	30 mb	10 mb	20 mb	18 mb
sex	f	m	m	f	f	m	m	m	m
child(c)/foetus(f)	f(33wg)	c(13m)	f(25wg)	f(24wg)	f(25wg)	f(26wg)	f(32wg)	f(23wg)	f(21wg)
iugr	+	nk	+	-	+	+	-	-	+
growth retardation	nk	+	nk	nk	nk	nk	nk	nk	nk
microcephaly	+	-	-	-	+	+	-	+	+
mental retardation	nk	+	nk	nk	nk	nk	nk	nk	nk
brain anomalies									
corpus callosum agenesis	-	-	+	-	+	nk	nk		
holoprosencephaly	-	-	-	-	+	+	+	+	
cerebellar vermis hypoplasia	+	-	+	-	-	nk	nk		

nk: not known, m: months, wg: weeks of gestation, f: female, m: male

Fig. 1 A (horizontal) table from the development set corpus, modified to fit the page

1   Contributions to Table Information Understanding

As an example of what needs to be done to extract information from a table, the table in Fig. 1 shows the three types of information: title, caption (or footnotes), and content. The content is contained in *cells*. The table information understanding problem can be seen as extracting and providing a semantics for this information.

Our contributions are a phenotype-genotype table ontology, a reading of the table cells that maintains relationships among the cells, and a tool that populates the ontology with the information extracted from the phenotype-genotype tables found in scholarly biomedical articles. Details of the first two contributions follow.

The Phenotype-Genotype Table Ontology. The process of building a phenotype-genotype table ontology required the curation of a sufficient number of texts containing a variety of tables in order to provide credence for our ontology. Using 107 tables found in 50 papers curated from our selected domain as our development set, we have engineered a table ontology that extends a subpart of the UMLS ontology with new concepts that are present in these tables (e.g. subjects have an age) as well as other fundamental concepts. This ontology, a portion is shown in Fig. 2, provides a semantics for the table data. In addition to storing the table data, the relationship among the cells (shown by the red lines), which is important information conveyed by the table structure can also be maintained.

Our ontology reproduces appropriate pieces of the UMLS ontology. These parts of our ontology are verified simply by the acceptance of the UMLS ontology. For concepts like genotype and phenotype, which are not in the UMLS ontology we have used an expert's knowledge to assist us. For example, we have added a concept named ORGANISM to accommodate certain important classes, like PATIENT and FAMILY, and certain important attributes for them, like AGE, GENDER, BIRTH WEIGHT, etc. To accommodate cell values that act as an identifier for the data in a column or row (e.g. PATIENTS, as in Fig. 1), we designed a generic concept named IDENTIFICATION ENTITY.

Table Information Understanding. To understand the table information, we begin with Hurst's concept of a reading path [2] to associate the *access cells* and the *data*

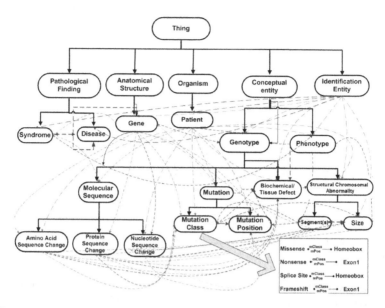

Fig. 2 A portion of the proposed ontology generated from the curated tables

cells. Data cells are characterized as cells where data appears in a terminating role. Access cells, which can be conceptualized as descriptors of the data cells, are the remaining cells. The reading of a table is the reading of all of the data cells.

Hurst views table data cells as independent entities. We realize that for the type of tables that are in our corpus, the data cells are connected (for instance, a column or row represents data about a single patient). These connections need to be maintained in the ontology in order to have a complete interpretation of the table. Hurst's concept of reading path is modified to maintain these relationships. In some cases to achieve the correct interpretation of the data cell values in our corpus, the primary relationship among the access cells may need to be changed from a vertical (column) relationship to a horizontal (row) relationship. To accomplish this, we change the order of the sequence of the reading path. For example, for Fig. 4, the primary headings are in columns and the data are distributed for the subjects down the columns. However, for Fig. 1, the primary headings are in the rows and the data are distributed across the rows. We refer to the previous column-oriented table as a vertical table and the latter row-oriented table as a horizontal table.

Heuristics are required to distinguish horizontal and vertical tables automatically. After observing our collection of tables, we designed the following heuristics: horizontal tables usually have many more columns than vertical tables; the first column heading starts with the Family or Patient ID followed by Age/Gender/Weight/Height attributes of the patient as row headings; and the remaining column headings contain attributes such as patient numbers or Identification numbers rather than alphabetic attributes (see Fig. 1). Very often, the data cell values are expressed as symbols.

Using the reading path concept we can reach a particular data cell. For example, considering the table in Fig. 4 the reading path for the data cell in row 2 column

access cell c_{00} (column/row heading 0)	access cell c_{01} (column heading 1)	access cell c_{02} (column heading 2)	$\cdots$	access cell c_{0n} (column heading n)
access cell c_{10} (row heading 1)	data cell c_{11}	data cell c_{12}	$\cdots$	data cell c_{1n}
$\cdots$				
$\cdots$				
$\cdots$				
access cell c_{m0} (row heading m)	data cell c_{m1}	data cell c_{m2}	$\cdots$	data cell c_{mn}

Fig. 3 Abstract table indicating access cells and data cells (m rows, n columns)

Table 1: HLXB9 Mutations Identified in the Study and Associated Phenotypes

Mutation Class	Mutation Position	Nucleotide Change	Amino Acid Change	Clinical Phenotype	Family or Patient No.
Missense	Homeobox	C→G, nt 4171	R247G	Hemisacrum, ARM, presacral mass, perianal abcess	3
Splice Site	Homeobox	A→G, nt 4889	NA	Hemisacrum, ARM, presacral mass, nonpenetrance	16
Frameshift	Exon 1	Ins C, nt 125-30	NA	Hemisacrum, ARM, presacral mass, neurogenic bladder, nonpenetrance	20

Fig. 4 Example of a vertical table (reduced in size) from the corpus of 115 phenotype-genotype tables

2 (containing value Homeobox) is: Mutation Class → Missense → Mutation Position → Homeobox. Using this simple reading path, we are able to insert the values Missense and Homeobox and the relation between them in the ontology. However, using the reading path concept for each data cell, it would not be possible to retain the relation between Homeobox and the other cells in the row. Instead, we combine all of the reading paths of all cells in a row, taking common terms only once.

According to our revised reading path, the reading path for the second row of the table in Fig. 3 is: cell c_{00} → cell c_{10} → cell c_{01} → cell c_{11} → cell c_{02} → cell c_{12} → …cell c_{0n} → cell c_{1n}. If we apply this to the table in Fig. 4 we get: Mutation Class → Missense → Mutation Position → Homeobox → Nucleotide Change → C→G, nt 4171 → amino acid change → R247G → Clinical Phenotype → Hemisacrum, ARM, presacral mass, perianal abcess → Family or Patient No. → 3.

Now we search for the first term MUTATION CLASS in the ontology and once we find it we insert the second value "Missense" under the class MUTATION CLASS. Similarly, we enter other concept-value pairs from the reading path. In this way we can populate our ontology appropriately. Moreover, from the reading path we know how data cells are connected with each other. We reflect this connection into our ontology by creating a relationship and connect data cells with it. As an example, to preserve the relationship between data cells under MUTATION CLASS and MU-TATION POSITION we build a relationship named "mClassmPos" and connect data cells with it, which is illustrated in Fig. 2.

For a horizontal table we need to change the order of the reading path: cell c_{00} → cell c_{01} → cell c_{10} → cell c_{11} → cell c_{02} → cell c_{12} → …cell c_{0n} → cell c_{1n}. Considering the horizontal table in Fig. 1, the reading path for row 3 is: Patients → 1 → Size of deletion → 70 Mb → 2 → 30 Mb → 3 → 47 Mb → 4 → 28 Mb → 5 → 34 Mb → 6 → 30 Mb → 7 → 10 Mb → 8 → 20 Mb → 9 → 18 Mb Now we can populate the ontology as we did for Fig. 4.

To solve the problem of missing column headings, we take the caption of the table as a source of the appropriate heading. To confirm this choice we check whether the values of that column fall under this concept. In the future, this will be performed automatically, but here, these two steps were performed manually.

We have observed column (or row) heading that have values like "Classes of Mutation" or "Position of Mutation", instead of directly matching the name of the concept in our ontology (Mutation Class and Mutation Postion). In these cases we check the possible word variations and further confirm our choice by observing the often highly stylized form of the data associated with that heading, making certain that it corresponds to the ontology concept that we have chosen.

2 Evaluation

The system correctly populates the ontology with the information contained in the development set of 107 genotype-phenotype tables.[1] The proposed ontology and population method are further verified by populating the ontology with the data from 31 previously unseen vertical tables, curated from 17 papers using the same keyword search, comprising 150 columns in total.[1] Column headings should map to concepts in the ontology. To calculate the accuracy of our system we consider the number of columns that successfully map into the table ontology, success being marked by the software finding a concept to map to and correctness of the concept being verified by human judgement. According to this criteria, the 120 correctly mapped columns gives the following accuracy: $\frac{120}{150} * 100\% = 80\%$. The causes of missed or incorrect interpretations for the 30 columns by our current system are summarized below.

Headings in 18 columns representing 10 distinct concepts do not map to a concept in our current ontology. In 5 cases the column header refers to a concept, but the values in the column belong to an aspect or property of the concept or a different concept. For 5 cases the column heading synonym list is inadequate for the mapping to the correct concept that exists in the ontology. In two other cases we found one column missing a column heading and one column heading which is actually a combination of two concepts joined by "and".

The first problem, finding concepts in tables that are not in our ontology, has been anticipated: the ontology is meant to evolve, especially in its gestation period. Most of the mapping problems encountered by our system will be overcome with appropriate updates to the ontology, which include having a good base of linguistic synonyms that map to the same ontological concept. We are currently investigating automatic and semi-automatic methods for adding concepts to the ontology. The second problem is much rarer. We already have procedures in place to confirm the mapping of column/row headers using the data values in the column/row (for instance, we do this for the missing column header problem).

---

[1] `http://www.csd.uwo.ca/~mercer/PhenGenTable-corpus-bibliography` provides a bibliography of the 67 papers and
`http://www.csd.uwo.ca/~mercer/PhenGenTable-corpus`, the corpus of 138 tables.

3 Related Work

Wong *et al.* [4] provides an automated system to extract information about mutations (gene, exon, mutation, codon and related statistics) from tables. They classify the table data to map the column/row values to a relevant entity, and then extract mutation information from these data. Mulwad *et al.* [3] introduces a domain independent framework for the intended semantics of tables. Column headers are mapped to class labels from an ontology; relationships between columns are discovered; cell values are linked to Linked Open Data entities and appropriate linked data.

In comparison, our work provides a domain-based ontology to store not only the data from the table but also the relationships that hold among the data cells. Furthermore, we are interested in a broader range of concepts for our ontology than the first work: mutation, gene, exon, phenotype, genotype, disease, syndrome.

4 Conclusions and Future Work

This paper reports on a table ontology designed to represent the tabular information in phenotype-genotype tables in scholarly biomedical papers. We extend the reading path concept to make it functional for our concept of table orientation, to populate the ontology with data and to preserve the various relationships among the table data. The populated ontology represents the semantics of each piece of information and preserves the relationship among cells in the table.

For future work, adding concepts automatically or semi-automatically to the incomplete ontology needs investigation. Unanticipated complications encountered during evaluation need to be addressed. As well, two issues arose that address aspects of the design and population of ontologies in a more general way. Firstly, we discovered in our evaluation phase, one table that referred to *the lack* of a mutation. Secondly, our biomedical expert has pointed out that knowledge changes over time and this is reflected in how the data is reported (e.g. epigenetic changes are not understood as well as sequence-based or structural chromosomal changes; and uncertainty in interpretation will be communicated in inconsistent ways). Our ontology will have to address this temporal aspect to record information of these types.

References

1. Hunter, L., Cohen, K.B.: Biomedical language processing: What's beyond PubMed? Molecular Cell 21(5), 589–594 (2006)
2. Hurst, M.F.: The interpretation of tables in texts. PhD thesis, University of Edinburgh (2000)
3. Mulwad, V., Finin, T., Joshi, A.: A domain independent framework for extracting linked semantic data from tables. In: Ceri, S., Brambilla, M. (eds.) Search Computing. LNCS, vol. 7538, pp. 16–33. Springer, Heidelberg (2012)
4. Wong, W., Martinez, D., Cavedon, L.: Extraction of named entities from tables in gene mutation literature. In: Workshop on BioNLP, pp. 46–54 (2009)

Part IV
Knowledge Representation and Reasoning

An Implementation of a Menu-List Recommendation System Providing Feedback from User

Chika Nishikawa, Akihiko Nagai, Takayuki Ito, and Satomi Maruyama

Abstract. People are increasingly searching for recipes online when they cook. At "cookpad" [1], Japan's most popular recipe site, users can search for and contribute recipes. But since such recipe sites often fail to provide detailed nutrition information, users have to determine balanced nutrition by themselves. Hence, our system, which recommends a menu-list based on nutritional balance and considers user feedback about its recommended menu-list. When users input what they actually have eaten, our system recommends meals based on the feedback information after considering the entire nutritional content. The experimental results suggest that our system can provide useful nutrition information. These results were validated by a nutritional expert.

1 Introduction

First, we focus on three works that are related to our research. One is a goal-oriented recipe recommendation system that utilizes nutritional information from the Internet [2]. Another is an automatic nutrient calculation system that determines how particular cooking methods change the nutrients of the foods used for preparing meals [3]. The last work is a menu recommendation system based on various food combinations that are eaten together [4]. Since our system recommends a one-week menu-list as meals composed of various menus and reflects user feedback about the menu-list recommendations, it differs from these related works.

Chika Nishikawa · Akihiko Nagai · Takayuki Ito
Gokiso-cho, Showa-ku, Nagoya-city, Aichi, Japan
e-mail: {nishikawa.chika,nagai}@itolab.nitech.ac.jp,
        ito.takayuki@nitech.ac.jp

Satomi Maruyama
1723, 2-chome,Omori, Moriyama-ku, Nagoya-city, Aichi, Japan
e-mail: maruyama@kinjo-u.ac.jp

M. Ali et al. (Eds.): *Contemporary Challenges & Solutions in Applied AI*, SCI 489, pp. 55–60.
DOI: 10.1007/978-3-319-00651-2_8   © Springer International Publishing Switzerland 2013

2 Menu-List Recommendation System

2.1 System Architecture

We show the architecture of our system in Fig. 1. The web server that receives the user's food information outputs the results that match the cooking information in the HTML using the databases. We match the information with the databases in the recipe recommendation engine, which calculates the amounts of nutrient, selects suitable meals, and outputs the results.

Our system uses two databases: ingredient/recipe and food. We have information about 8.6 million ingredients and 1.6 million recipes. In the ingredient/recipe database, we have each data of the quantity of ingredients and ingredients included in in each recipe. It unifies the amounts of food, one serving to select meals efficiently. The food database, which has the amount of energy of such processed foods as a hamburger,is based on the average amount of energy information about 2500[5]. Our system refers to such information to determine the amount of energy when users input their feedback information.author

2.2 System Overview

First, users input the health and any food allergies information. The system checks the database to match this information and calculates the nutrient amounts. Then it selects recipes that meet the user requirements. Next, it can adjust the recommended menus not only for one meal but the entire nutritional balance. Then our system generates a one-week menu-list on the web. After generating the menu-list, the user inputs the food that he actually eats as feedback. Finally, our system adjusts the meals and generates the menu-list reflecting feedback.

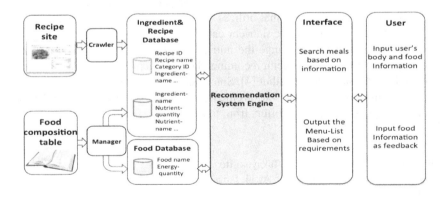

Fig. 1 System architecture

3   Menu-List Recommendation Mechanism

3.1   User Information

First, users login when they use the system. They input their health requirements and any foods to which they are allergic.

Amount of Energy: The amount of daily required energy determines their requirements based on the basal metabolic rate (BMR). We use Harris-Benedict Equation for BMR [6]. This equation differs calculation formula by men and women. The basal metabolic rate is comprised of 70% of the total amount of energy consumed in a day, as defined by a Japanese government health agency. The default maximum of the amount is determined with this rate if users don't input it.

Food Allergies: Users input the ingredients/foods to which they are allergic in the user information to confirm whether such foods are included in the meals.

3.2   Generating a Menu-List

3.2.1   Daily Requirement of Each Nutrient

Our system takes into account the amount of energy, cholesterol, and sodium of menus in one day: energy is a measure of the nutrient balance of the whole, cholesterol and sodium are closely related to lifestyle-related diseases[7]. Each nutrient amount is based on standard daily recommended allowances; energy,determined by BMR, 500 to 600 mg of cholesterol, and a maximum of 9 grams to sodium.

3.2.2   Determining Meals

First, a "meal" is composed of breakfast, lunch, and dinner. We differentiated between breakfast and the other meals.

In breakfast, we choose menus that can be cooked easily to ease the burden on the stomach[8]. Then we collected recipe pages that included keywords morning and breakfast. In lunch and dinner, this system separately recommends main and sub dishes [9]. Main dishes are composed of ingredients that include many proteins, fat, energy, and iron, such as meat, fish, eggs, and soybeans. Sub dishes are composed of ingredients that include lots of vitamins, and minerals like seaweed, vegetables.

Fig. 2 shows the flow to determine one meal in lunch and dinner. First, This system decides the condition of generating the menu-list based on the user's information. Next, it selects the menus which may be a candidate for main dish using the database. This system checks whether the selected menu satisfies the energy condition. If this menu doesn't satisfy it, this system selects the main dish again. If this menu satisfies it, this system selects the menus which may be a candidate for sub dish using the database. This system also checks the sub dish in the energy condition. The system adds the sub dishes as the total menus satisfy that more than BMR and less than the upper limit of the energy. So, our system recommends one main dish and some sub dishes.

3.2.3 Example

Fig. 3 shows an example of the system interface. This user is a 30-year old, 160-centimeter tall woman who weighs 50 kilograms. A menu-list shows the recipe names and thumbnails for them.

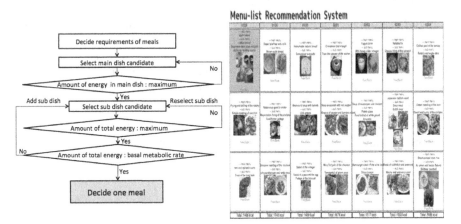

Fig. 2 Flow for recommending one meal   **Fig. 3** User Interface

4 User Feedback

Our system can get the food information that the user actually eats as feedback. By matching the food database, we roughly determine the amount of energy and adjust meals based on it. We propose the following three ways to use the feedback by focusing on the energy requirements. Energy requirements are the relation between the total energy and, BMR and the upper limit of energy. When the total energy satisfies more than BMR and less than the upper limit after feedback, our system changes from the recommended food to that given in the feedback. Our system changes only the feedback food because it is not necessary to change the other recommended meals. When the total energy is less than BMR, our system adjusts them and adds a sub dish. For example, for breakfast, the user inputs yogurt as feedback, which has few calories. Our system adds a sub dish at dinner and adjusts the nutritional balance. When the total energy energy is more than the upper limit, this system has the following two ways to adjust the entire nutritional balance. First, our system deletes the recommended meals. By deleting meals, the whole energy is decreased, so our system adjusts the whole balance. When it cannot adjust the balance by deleting meals, it changes two meals. It again recommends new menus with few calories and adjust the amount. For example, for breakfast, the user inputs pasta as feedback, which has many calories. The system changes meals of lunch and dinner, and adjusts the whole balance. Fig. 4 and 5 shows an example of adjusting feedback.

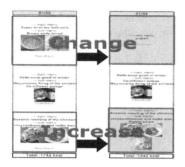 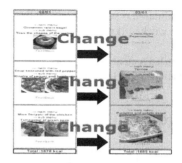

Fig. 4 Feedback (when not enough energy) **Fig. 5** Feedback (when too much energy)

5 Performance Evaluation of System

The nutritionists evaluated the results using the protein, fat, and carbohydrate (PFC) ratio by calculating the each recommended meals for two weeks. We can know the nutritional balance by evaluating the big three nutrient. The ideal protein ratio is 10-20%, fat ratio is 20-25%, and carbohydrate ratio is 50-70%.

Evaluation Result: The evaluated protein ratio was 16.8& and carbohydrate ratio was 55.4%, which satisfy each ideal percentage. But, the evaluated fat ratio was 27.8%. It is caused by the recommended foods included the same ingredients that have many fat in one day (46.2%) and the average fat rate is risen. So, we have to improve the selection of meals considering overlap of ingredients.

6 Related Work

Research on systems that recommend recipes has gained attention.

Ueda et al. proposed a recipe recommendation system. This system assesses likes and dislikes by analyzing the history whether a user cooks or not and whether a user uses recipes.[10]. However, our system recommends well balanced meals.

Karikome et al. proposed a system that can support eating habits [11]. This system can search recipes, allow a user to maintain a dairy food, and recommend recipes. This system recommends meals based on nutritional information and the user's diary. However, our system takes user feedback and adjusts the menu.

Gijs et al. proposed a system that recommends a personalized recipe, which facilitates its users to make health-aware meal choices based on past selections[12]. It recommends well-balanced meals, and represents the nutritional balance. However, our system takes user feedback and user's health information for deciding BMR.

7 Conclusion and Future Work

We proposed a system that recommends a menu-list based on nutritional balance. Our system adjusts the meals based on feedback as to well-balanced nutrition. From the evaluation result, we confirm that our system recommends well-balanced meals. Future work will recommend staples food, and consider the overlap of ingredients.

Acknowledgements. This work is partially supported by the Funding Program for Next Generation World-Leading Researchers (NEXT Program) of the Japan Cabinet Office.

References

1. Cookpad (2013), `http://cookpad.com`
2. Ueta, T., Iwakami, M., Ito, T.: Implementation of a Goal-Oriented Recipe Recommendation System Providing Nutrition Information. In: TAAI (2011)
3. Takahashi, J., Ueda, T., Nishikawa, C., Ito, T., Nagai, A.: Implementation of automatic nutrient calculation system for cooking recipes based on text analysis. In: Anthony, P., Ishizuka, M., Lukose, D. (eds.) PRICAI 2012. LNCS (LNAI), vol. 7458, pp. 789–794. Springer, Heidelberg (2012)
4. Takahashi, J., Ueta, T., Ito, T.: A Menu Recommend System based on being eaten together. JAWS (2011) (in Japanese)
5. e-kenko.co.jp (2013), `http://www.ekenko.co.jp/`
6. foundation and structure of diet, `http://diet.shining-eternally.com/`
7. health clinic (2013), `http://www.health.ne.jp`
8. Morning-Time.jp (2013), `http://www.asajikan.jp/asagohan`
9. Asken (2013), `http://www.asken.jp`
10. Ueda, M., Takahata, M., Nakajima, S.: User's Food Preference Extraction for Personalized Cooking Recipe Recommendation. In: ISWC (2011)
11. Karikome, S., Fujii, A.: A System for Supporting Dietary Habits: Planning Menus and Visualizing Nutritional Intake Balance. In: ICUIMC, pp. 386–391 (2010)
12. Geleijnse, G., Nachtigall, P., van Kaam, P., Wijgergangs, L.: A Personalized Recipe Advice System to Promote Healthful Choices, IUI, pp. 437–438 (2011)
13. Nishikawa, C., Nagai, A., Ito, T.: An Implementation of a Menu-List Recommendation System Providing Feedback from User. JAWS (2012) (in Japanese)

On Representing and Sharing Knowledge in Collaborative Problem Solving

Heba Al-Juaidy, Lina Abu Jaradeh, Duha Qutaishat, and Nadim Obeid

Abstract. In this paper, we propose an agent-based framework for collaborative problem-solving. We emphasize the knowledge representation and knowledge sharing issues. We employ a three-valued based Temporal First-Order Nonmonotonic Logic that allows an explicit representation of   events/actions and can handle dialogue game protocols and temporal aspects explicitly. A prototype is developed with a case to guide and assist evacuees in an emergency evacuation from a building.

Keywords: Knowledge Sharing, Multi-Agent Systems, Knowledge Representation, Problem Solving.

1   Introduction

Collaborative Problem Solving (CPS) is the process by which a collection of intelligent agents work together to partition a complex, large and/or unpredictable problem into an appropriate set of simpler sub-problems where each will be (partially) solved by one or a group of expert agents and finally the partial solutions are integrated to produce a solution to the whole problem[14]. This decomposition allows each agent to use the most appropriate technique to solve the sub-problem to which it is assigned. Multi-Agent System (MAS) represents an appropriate approach for solving inherently distributed problems, whereby clearly different and independent processes can be distinguished. In CPS settings, the use of MAS offers conceptual clarity, flexibility, the ability to handle applications with a natural spatial distribution and with uncertain information.

Heba Al-juaidy · Lina Abu Jaradeh · Duha Qutaishat · Nadim Obeid
Department of Computer Information Systems, King Abdullah II School for Information Technology, The University of Jordan
email: obein@ju.edu.jo

M. Ali et al. (Eds.): *Contemporary Challenges & Solutions in Applied AI*, SCI 489, pp. 61–66.
DOI: 10.1007/978-3-319-00651-2_9   © Springer International Publishing Switzerland 2013

In this paper, we propose an agent-based framework for CPS. We emphasize the Knowledge Representation (KR) and Knowledge Sharing (KS) issues in a distributed agent-based system. We employ a three-valued based Temporal First-Order Nonmonotonic Logic (TFONL) that allows an explicit representation of events/actions [cf. 13, 16] and can handle dialogue game protocols and temporal aspects explicitly. A prototype is developed with a case to guide and assist evacuees in a emergency evacuation from building. In section 2 we discuss the suitability of MAS for CPS. In section 3 we discuss KR and KS issues and present the example. Section 4 is dedicated to discussions and related works.

2   CPS and MAS

The idea of CPS is to decompose a problem into a set of sub-tasks where each has some form of relation/association with other sub-tasks that must be dealt with by the appropriate agents that possess enough PS knowledge to apply their own expertise to its sub-task. Decentralization of the tasks seems to be a reasonable way to keep control within large complex problems [8]. Multi-Agents Systems (MAS) have been proposed as a suitable model for handling complex, distributed and heterogeneous systems [7, 15]. An MAS can be defined as: a collection of agents with their own problem solving capabilities and which are able to interact among them in order to reach an overall goal [8]. Agents are specialized problem solving entities. They are autonomous as they have control both over their internal state and over their actions.

In CPS Situation, agents can help each other by negotiating a partition of the problem into manageable sub-problems/tasks among themselves according to their abilities, expertise and skills. The key issue to be resolved in sub-problems and their associated tasks is how tasks are to be distributed among the agents and how to dynamically configure the system by distributing the software components on the available hardware hosts so that they work together as a whole unit to meet changing requirements [1].

Due to the dynamic and unpredictable nature of the environment, in which agents operate, it is not possible to give, at the outset, a complete specification of all the tasks and the knowledge/expertise required. Thus, there is a need for an environment that integrates the knowledge of the various agents and partial results, so that agents could have access to information, expertise and knowledge they need. Depending on its knowledge and reasoning ability, an agent may pursue an objective/goal in order to help it in deciding what method and/or technique to use for another objective/goal. A CPS task may encompass planning, scheduling, and collaborative diagnosis [7]. Collaboration and different expertise has its problems. CPS may require the collaborators to be involved in a process of reformulation and questioning until they reach a point of consensus; we deviate from one complete solution and rather develop solutions in small steps refinements [6].

3   Dialogue Moves in TFONL

In CPS situations, agents have incomplete knowledge of their environments. This makes reasoning complex since the closed world assumption can no longer be applied; an agent cannot assume that a fact is false just because it does not know about it. Therefore, it is important to cater for representing and updating the incomplete knowledge which would be useful to an autonomous agent capable of KS and of manipulating its environment. It is possible to express an agent's knowledge using classical logic. However, it may prove to be more difficult and quite unnatural. Furthermore, agents in a dialogue make statements that are realistic based on the context. However, these statements may become futile during the dialogue when information previously unknown become available which may cause agents to revise their knowledge. Defeasible logic is suitable for the dynamics of argumentation and dialogue where agents could change their beliefs.

In this paper, we employ a three-valued based Temporal First-Order Nonmonotonic Logic (TFONL) that allows an explicit representation of time   and events/actions [cf. 13, 16]. TFONL is an extension of the quantified version of the non-temporal system T3 [cf. 10, 11, 12]. The language, L_{T3}, of T3 is that of Kleene's three-valued logic extended with the modal operators "M" (Epistemic Possibility) and "P" (Plausibility). The material implication "⊃" can be defined as follows: $(A \supset B = M(\sim A \,\&\, B) \lor \sim A \lor B$. In T3, "L" is the dual of "M" and "N" be the dual of "P", i.e., $LA \equiv \sim M \sim A$ and $NA \equiv \sim P \sim A$ where $A \equiv B$ is equivalent to $A \supset B$ and $B \supset A$. $A \Rightarrow B$ represents the default A A:B/B [10].

Nonmonotonic reasoning is represented via the operators M (*epistemic possibility*) and P (*plausibility*). Informally, MA states that A is not established as false. Using M, we may define the operators U (*undefined*), D (*defined*) and ¬ (*classical negation*) where UA is true if the truth value of A is undefined and DA is true if the truth value of A is not undefined. More specifically, $UA \equiv MA \,\&\, M \sim A$, $DA \equiv \sim UA$ and $\neg A \equiv DA \,\&\, \sim A$.

Within the framework of TFONL, it is possible to formalize dialogue moves and the rules of protocols of the required types of dialogue. These rules are nonmonotonic because the set of propositions to which an agent is committed and the validity of moves vary from one move to another. Let L_{Com} specify the locutions which the agents participating in a dialogue are able to express. A dialogue consists of a course of successive moves made by the participants. A Dialogue Move can be defined as follows:

Definition 3.1. A Dialogue Move M can be defined as a 7-tuple as follows:

M=<Id(M), Sender(M), τ(M), δ(M), Content(M),Receiver(M), Target(M)> where Id(M) is the identifier of M, Sender(M) is the speaker of <δ(M),Content(M)>, τ (M) is the time of M, δ(M) ∈ {Assert, Accept, Reject, Retract, Question, Justify, Challenge }, Content(M) is the content of M, Receiver(M) is the addressee and Target(M) is a previous move to which M is a reply.

We now present our example. Emergency evacuation is the urgent movement of people from a place due to the occurrence of some dangerous event [2] which is challenging. MAS are particularly suitable for assisting in such tasks.

The Computer Information System (CIS) department in at the university of Jordan consists of seven computer labs, five lecture halls, a students' hall and service rooms. Let A_1, ..., A_k stand for areas, Z_1, ..., Z_r for Zones. We shall use Z_{ij} to denote zone i in area j. The zones in CIS are as follows: Z_{11} = [Exit1 ,Lab206 ,Lab207 , k205], Z_{21} = [2 Elevators, Lab201 ,K204 ,k201 , sitting room], Z_{12} = [Exit2, Service rooms, Lab203, Lab202, K202] and Z_{22} = [Exit3 , Lab 204, Lab 205, K 203] (cf. Fig. 1 and Fig. 2).

With each zone Z_{ij}, we associate an agent group G_{ij} that includes Z_{ij}-Supervisor, Z_{ij}-Monitor, Z_{ij}-Info-Coll, Z_{ij}-Guide. In case of an anomaly in a zone Z_{kl}, the Z_{kl}-Monitor informs all agents in its zone namely, Z_{kl}-Supervisor, Z_{kl}-Info-Coll and Z_{kl}-Guide. The Z_{kl}-Supervisor informs the Al-Supervisor of the anomaly in order to activate the appropriate alarms and/or call rescue teams. It keeps track of the workflow of the evacuation process and manages the actions performed by each agent in its zone. It can negotiate with other zones' supervisors if it needs help to carry out its plans such as guiding people through appropriate zones to avoid congestion. Z_{kl}-Info-Coll collects information about the state of its zone and other zones such as congestion, safety, open/closed gates and the states of evacuees such as injuries, agents breakdown and so on. Z_{kl}-Guide guide people to safe exits using safe paths taking into consideration what is reported by Z_{kl}-Info-Coll.

With each area A_l, we associate an agent group G_l that includes A_l-Supervisor, A_l-Info-Coll, A_l-Planner, A_l-Guide. The A_l-Supervisor, A_l-Info-Coll and A_l-Guide have similar tasks to those at zone levels at the area level. The A_l-Planner determines the alternative sets of safe routes through the different zones in A_l and other areas to an exit and send it to A_l-Guide.

The agents can make use of fluents such as gas-smell, fire-heat, smoke, emergency and so on. The environment may include propositions such as Exit(Gate1), Exit(Gate2), Location(Gate1, Z11), Location(Elevator1, Z21), State(Gate2, Closed) and State(Gate1, Open). We employ rules such as (R1) and (R2):

(R1) T $\Rightarrow$ Clear(P) &Path(P)

(R2) Exit(G)&Location(G, Zij)&on-Path(P, G)&State(G, Closed)&
$$\neg Crowded(G) \Rightarrow Open(Zij\text{-}Sup, G).$$

(R1) states that by default, paths are clear. (R2) states that if there is an exit G that is not crowded, on a safe path, and G is closed, then open G.

Suppose that there is a fire in Z11 (cf. Fig. 1), then Z11-Monitor detects the fire and informs Z11-Supervisor. This move is not a reply to any previous move.

M_1 = <1, Z11-Monitor, t_1, Assert, fire in Z11, Z11-Supervisor, 0>

In M_2, Z11-Supervisor informs Z11-Monitor that it accepts the message.

M_2 = <1, Z11-supervisor, t_2, Accept, fire in Z11, Z11-Monitor, 1>

Z11-Supervisor informs A1-supervisor that there is a fire in Z11.

M_3 = <3, Z11-Supervisor, t_3, Assert, fire in Z11, A1-Supervisor, 0>

In M_4, A1-Supervisor informs Z11-Supervisor that it accepts the message.

M_4 = <4, A1-Supervisor, t_4, Accept, fire in Z11, Z11-Supervisor, 3>

A1-Supervisor can inform A1-Planner that there is a fire in Z11 and so on.

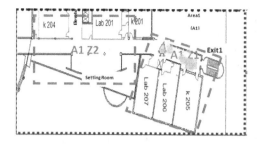

Fig. 1 Fire in Z11

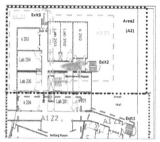

Fig. 2 Fire in Z12

In the same way, A1-Supervisor informs Zj1-supervisor, Ak-supervisors, and A1-Planner. A1-Planner ids required to propose an evacuation route.

Similarly, A2-Supervisor and each zone supervisor in A2 will inform appropriate agents do what is required. Each zone planner will propose a safe route through its zone. The area planners, will integrate these partial solutions to propose safe routes. A1-Planner has to provide a safe route and it may propose

R1: $Z11 \rightarrow Z21 \rightarrow$ELEVATORS and R2: $Z11 \rightarrow Z21 \rightarrow Z22 \rightarrow$Exit3 as in:

M_i = <4, A1-Planner, t_i, Assert , {R1, R2}, A1-Supervisor, 5>

Similarly, if there is a fire in Z12 (cf Figure 2) then the following steps may be taken: Z12-Monitor informs Z12-Supervisor of event. Z12-Supervisor informs A2-supervisor and all agents in Z12 of the event. A2-Supervisor informs Zj1-supervisor, Ak-supervisors and A2-Planner to propose an evacuation route. Each Ak-Supervisor inform agents in its area to do what is required according to their specialization. For instance, Z22-Planner proposes a route through Exit 3. Z21-Planner proposes a route through Exit 1. A2-Planner proposes a route through Exit 3. A1-Planner proposes a route through Exit 1. A1-Supervisor and A2-Supervisor can negotiate a plan to guide people.

4   Discussion and Comparison with Related Works

To our knowledge, little consideration is to dialogue and argumentation in CPS. In [7] the role of dialogue in MAS is shown. In [15] the role of agents in the development of a KS system is highlighted. Some formalisms have been suggested for specifying and verifying protocols or tracking agents' commitments during dialogue [3, 4, 9]. In [5], a generic framework for specification of dialogue game protocols is presented These protocols are based on classical logic. These approaches do not commit to a mechanism for agents to think about the acceptability of arguments. TFONL, however, can handle dialogue game protocols and temporal aspects explicitly.

We have in this paper emphasized the KR and KS issues. A prototype is developedd with a case to guide evacuees in a emergency evacuation form a building.

It is important to identify relevant commitments which an agent has to satisfy and to investigate how to integrate this proposal with techniques used in planning to identify strategies to satisfy important commitments [9]. It is useful to further investigate strategic and tactic reasoning in solving more complex problems.

References

1. Al-Areqi, S., Hudaib, A., Obeid, N.: Improving Availability in Distributed Component-Based Systems via Replication. In: Nguyen, N.T., Trawiński, B., Jung, J.J. (eds.) New Challenges for Intelligent Information and Database Systems. SCI, vol. 351, pp. 43–52. Springer, Heidelberg (2011)
2. Bonabeau, E.: Agent-Based Modeling: Methods and Techniques for simulating human systems. Proc. National Academy of Sciences USA 99, 7280–7287 (2002)
3. Chesani, F., Mello, P., Montali, M., Torroni, P.: Commitment Tracking via the Reactive Event Calculus. In: IJCAI, pp. 91–96 (2009)
4. Giordano, L., Alberto, M., Camilla, S.: Specifying and Verifying Interaction Protocols in a Temporal Action Logic. J. Applied Logic 5(2), 214–234 (2007)
5. McBurney, P., Parsons, S.: Dialogue Games for Agent Argumentation. In: Rahwan, I., Simari, G.R. (eds.) Argumentation in Artificial Intelligence, vol. 261 (2009)
6. Moubaiddin, A., Obeid, N.: The Role of Dialogue in Remote Diagnostics. In: 20th Int. Conf. on COMADEM (2007)
7. Moubaiddin, A., Obeid, N.: Dialogue and Argumentation in Multi-Agent Diagnosis. In: Nguyen, N.T., Katarzyniak, R. (eds.) New Chall. in Appl. Intel. Tech. SCI, vol. 134, pp. 13–22. Springer, Heidelberg (2008)
8. Moubaiddin, A., Obeid, N.: Partial Information Basis for Agent-Based Collaborative Dialogue. Applied Intelligence 30(2), 142–167 (2009)
9. Moubaiddin, A., Obeid, N.: On Formalizing Social Commitments in Dialogue and Argumentation Models Using Temporal Defeasible Logic. Knowledge and Information Systems (2012), doi:10.1007/s10115-012-0578-6
10. Obeid, N.: Three Valued Logic and Nonmonotonic Reasoning. Computers and Artificial Intelligence 15(6), 509–530 (1996)
11. Obeid, N.: Towards a Model of Learning Through Communication. Knowledge and Information Systems 2, 498–508 (2000)
12. Obeid, N.: Fault Diagnosis Using Three-Valued Based Nonmonotonic Logic. International Journal of COMADEM 3(2), 17–28 (2000)
13. Obeid, N.: A Formalism for Representing and Reasoning with Temporal Information, Event and Change. Applied Intelligence 23(2), 109–119 (2005)
14. Obeid, N., Moubaiddin, A.: On the Role of Dialogue and Argumentation In Collaborative Problem Solving. In: ISDA, pp. 1202–1208 (2009)
15. Obeid, N., Moubaiddin, A.: Towards a Formal Model of Knowledge Sharing in Complex Systems. In: Szczerbicki, E., Nguyen, N.T. (eds.) Smart Information and Knowledge Management. SCI, vol. 260, pp. 53–82. Springer, Heidelberg (2010)
16. Obeid, N., Rao, B.K.N.R.: On Integrating Event Definition and Event Detection. Knowledge and Information Systems 22(2), 129–158 (2010)

Part V
Machine Learning Applications

Constructing Language Models for Spoken Dialogue Systems from Keyword Set

Kazunori Komatani, Shojiro Mori, and Satoshi Sato

Abstract. Spoken dialogue systems (SDSs) need language models (LMs) for automatic speech recognizers (ASRs) for each domain. This is because domain-specific words such as proper nouns differ in domains, and they must be recognized correctly to accomplish the task. We propose a method to construct a class N-gram LM only from a set of domain-specific words (i.e., words in target relational database for retrieval). This problem setting corresponds to a situation where we construct a new spoken dialogue system; i.e., there is no sufficient corpus available in the target domain. We use a similar-domain corpus and assign class labels to it using machine learning. Because no sufficient training data are available, we create an initial training corpus by string matching and then use it as training data. The experimental results showed that our approach is promising: ASR accuracy for domain-specific words improved.

1 Introduction

Domain-specific words must be correctly recognized in spoken dialogue systems (SDSs) because they are requisite to accomplish their task. Here, the domain-specific words include place names, restaurant names, and other names in the relational database (DB) that is the target of retrieval. The requirements for constructing a language model (LM) for an SDS that retrieves a relational DB are as follows:

1. All domain-specific words in the DB need to be in its dictionary for automatic speech recognition (ASR). Such words in the same class (having the same attribute in the DB) can appear in the same context. Not every domain-specific word, however, necessarily appears frequently in a training corpus. This is because proper nouns in a specific domain are not common.

Kazunori Komatani · Shojiro Mori · Satoshi Sato
Nagoya University, Furo-cho, Chikusa-ku, Nagoya, Aichi 464-8603, Japan
e-mail: {komatani,s_mori,ssato}@nuee.nagoya-u.ac.jp

M. Ali et al. (Eds.): *Contemporary Challenges & Solutions in Applied AI*, SCI 489, pp. 69–76.
DOI: 10.1007/978-3-319-00651-2_10   © Springer International Publishing Switzerland 2013

2. The LMs need to be constructed easily. They are required for each domain and
   the contents of the target DB can be changed or updated.

One solution to satisfy these requirements, when no sufficient domain-specific corpus is available, is to use a class N-gram LM. In this LM, domain-specific words
are substituted into classes assigned to words in a corpus [1]. To construct the LM
simply, a large-scale corpus with classes is required; however, it is not realistic to
assume that such a corpus of the target domain is always available.

We construct a class N-gram model by exploiting a large-scale corpus of a similar
domain. A similar-domain corpus contains various utterance patterns but does not
contain all of domain-specific words; this is why a word N-gram LM does not work
for this purpose. We use a publicly available corpus as the similar domain corpus;
in general, it can be obtained from the Web after applying filtering measures, such
as the BLEU score [2], the relative entropy [3], and word perplexity [4, 5, 6]. We
only use a publicly available corpus in this paper because Web crawling is not our
main focus. The corpus is used as training data for the class N-gram LM, instead of
a target domain corpus, after class information is given to it by machine learning.

The main problem we address in this paper is that the training data for the machine learning are insufficient. No class information is given to the similar-domain
corpus, and an insufficient quantity of training data with domain-specific class information is available when an SDS for a new domain (i.e., DB) is constructed. We
try to solve this problem in a bootstrapping manner. That is, we assign classes to
the corpus by using a simple method (string matching) as the first step, and repeat
the training and "assigning labels" phases several times. We then construct a class
N-gram LM for the ASR by using the corpus to which class labels are assigned.

Some studies attempt to incorporate words in the target domain into a corpus
of the same style [7, 8]. This approach is similar to ours. One study is based on
predicate-argument (P-A) structures [7], and another uses similarity of words in
large-scale texts [8]. Our study focuses on the situation when only a target DB for
retrieval and a similar-domain corpus are available. This corresponds to the case
when a new SDS is constructed, that is, when available resources are more limited.

2   Problem Setting and Overview

The target of this research is to construct a class N-gram LM for SDSs that retrieve
a DB. Figure 1 shows the relationship between the LM and DB, and how the constructed LM will be used in an SDS.

We presuppose the following three resources are available to construct the LM.
This corresponds to a situation when a new SDS is constructed for a target DB.

1. Target DB: relational DB used as the retrieval target; that is, the system responds
   on the basis of the retrieval result obtained from this DB. Classes are defined by
   its attributes and domain-specific words are their values.

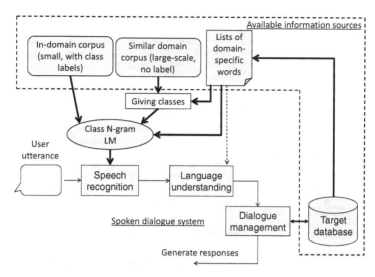

Fig. 1 How class N-gram LM will be used in SDS

2. Small corpus of the target domain: manually prepared by a system developer, consisting of several hundred utterances; however, this does not cover users' various utterance patterns and is not enough to be used as training data of machine learning.
3. Large-scale corpus of similar domain: a large-scale publicly available corpus; a corpus collected from the Web also can be used as this. It should contain more than several hundred thousand sentences and is thus expected to contain various utterance patterns. It does not necessarily contain domain-specific words in the target DB because its domain is only "similar." No class information is given.

Our challenge is how to give class labels to appropriate words in the similar domain corpus. This is requisite for constructing class N-gram LM because no sufficient amount of in-domain corpus with class labels is available, as explained in the presupposition. An overview of our method is depicted in Figure 2. As the whole, we give class information to the similar-domain corpus by a machine learning classifier. Since no training data is available, we make it by simple string matching with domain-specific words. Then, we construct a class N-gram LM with the similar-domain corpus to which class information was given.

Details of the procedure are as follows. The numbers correspond to those in the figure.

1. Classes and initial lists of domain-specific words are prepared from the target DB. Class information is derived from its attributes. The values of each attribute are stored in domain-specific word lists.
2. Classes are given to words that appear both in the domain-specific word lists and in the corpus. We call this process "string matching". The classes are defined as attributes of the target DB.

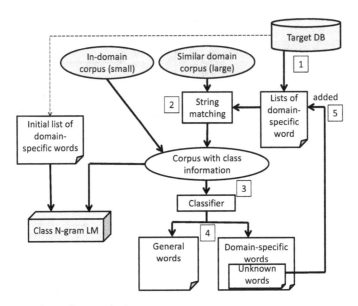

Fig. 2 Process flow of our method

3. A classifier is trained with the corpus obtained in Step 2.
4. All words in the corpus are classified into either a domain-specific word of each class or a general word, by the classifier; i.e., the same corpus is used both for training and classification. Then, unknown domain-specific words are obtained. Here, "unknown" means the word was not in the list of domain-specific words.
5. The unknown domain-specific words are added to the lists that will be used for the next string matching.
6. Return to Step 2.

By repeating this cycle, the numbers of domain-specific words in the lists will increase; this means that classes will be given to more words to which no classes were initially given. This will work to add contextual information around domain-specific words to the class N-gram LM (i.e., give connection information before and after domain-specific words).

3   Experiment on Giving Classes to Similar-Domain Corpus

Each word is classified into either the domain-specific word or the general one by the classifier, as described in Step 4 in the previous section. If a word is classified into a domain-specific word in a class, it can be regarded as assigning the class label to the word. Here, we use the maximum entropy model [9] based classifier.

We adjust the infrequency of the domain-specific words by introducing parameter t. Few class labels are contained in the similar-domain corpus after the string matching. This is because the corpus is "similar" to the target domain, and thus,

Table 1 Classes in target DB

Class names	Examples of domain-specific words	No. of words
FOOD	Chinese noodle, Pasta	83
GENRE	Chinese, Japanese	23
LOCATION	Sakae, Kita-ward	20
STATION	Motoyama, Nagoya	231

Table 2 Results of extracting domain-specific words ($N = 4$)

t	Number of unknown domain-specific words		Appropriate
	Appropriate	Not Appropriate	ratio
1	1	2	0.33
2	7	4	0.64
3	16	6	0.72
4	23	7	0.77
5	25	8	0.76
6	25	10	0.71
10	28	11	0.72
20	37	14	0.73

domain-specific words in the DB do not appear in it frequently. More specifically, we multiply the occurrence of sentences including all the domain-specific words by t and then train the classifier.

As feature sets of the maximum entropy model, we use N words before and after the target word. For example, if we use four words before and after the word ($N = 4$), one 4-gram, two 3-grams, three 2-grams, and four 1-grams are used to represent the context for both before and after the target word. We also use part-of-speech information of the target word. The value of N was determined after preliminary experiments.

The experimental setting follows. The target domain is a restaurant search in Aichi prefecture, Japan. We prepared the DB for it, which consists of 2398 entries. The domain has 78 attributes, and we selected 4 attributes from them. Thus, the classification is performed among five classes; that is, these four classes and one other (null). The four attributes and their examples are listed in Table 1. As for the similar-domain corpus, we used ten thousand sentences in the question sections of the "Cook, gourmet, and recipe" category of "Yahoo! Chiebukuro data (2nd edition)." We use Mecab [10] for word segmentation and Mallet [11] for the maximum entropy model.

We confirm whether any unknown domain-specific words were extracted by our method. First, we check whether the number of extracted domain-specific words can be controlled by changing the parameter t. Table 2 lists the number of extracted domain-specific words. One of the authors manually judged whether they were appropriate; that is, whether an extracted word is in the corresponding class or not.

Table 2 indicates that when a larger t was used, more domain-specific words were extracted. The ratio, which is the number of appropriate words to the total number

Table 3 Constructed LMs for comparison

LMs	Words used in giving classes (i.e., for string matching)
#1 (word N-gram)	None
#2 (only string matching)	DSWs in target DB
#3 (proposed)	DSWs in target DB + DSWs (automatically obtained)
#4 (proposed + selection)	DSWs in target DB + DSWs (manually selected)

DSW: Domain-specific word

of unknown words, gradually increased and almost leveled off when a larger t was used. Improving the ratio is desirable in order to obtain correct class labels, by using other classifiers or by changing the feature sets, for example.

4 ASR Using Constructed Class N-gram LMs

We confirm whether the proposed method can improve the ASR accuracy; more specifically, whether the class labels given in the previous section are helpful for ASR.

We compared the four LMs listed below. These LMs are summarized in Table 3.

LM #1: word N-gram LM trained with the similar-domain corpus; this does not use any class information. The vocabulary size is 33178.

LM #2: class N-gram LM whose training data are compiled in the first string matching only; that is, no classifier is used. The vocabulary size is 33071.

LM #3: 33 extracted words were also used for the string matching in addition to LM #2. This is our proposed model. The vocabulary size is 32806.

LM #4: Only 25 appropriate domain-specific words were manually selected from those used in LM #3. The vocabulary size is 32902. This objective of this LM is to guess the accuracy when the classifier in the previous section improves.

For the similar-domain corpus here, we used a larger test set: about 1,200,000 sentences in the "Cook, gourmet, and recipe" category of "Yahoo! Chiebukuro data (2nd edition)." For a small domain corpus, we collected 132 sentences by hand, multiplied their occurrences 10000 times, and mixed them into the similar-domain corpus. The mixed corpus was used to construct the LMs. We used Mecab for text segmentation as well as the previous section, and Julius[1] as the speech recognizer. The classes were the same as listed in Table 1. We set the uniform probabilities inside the classes when constructing class N-gram LM.

For the evaluation set, we used another utterance set that had been collected by using a spoken dialogue system that retrieved the same restaurant DB. There were 120 dialogues (30 subjects performed 4 dialogues each). We used 4480 utterances; very short mutters and noises were removed. The utterances contained 14554 words including 1454 domain-specific words. The criteria we used was word accuracy (Acc.) for all and domain-specific words.

---

[1] http://julius.sourceforge.jp

Table 4 ASR accuracy for each LM

LMs	All words	Domain-specific words
#1 (word N-gram)	64.22	73.5
#2 (only string matching)	64.04	73.9
#3 (proposed)	63.81	75.0
#4 (proposed + selection)	64.10	74.1

$$Acc. = (N - Del - Sub - Ins)/N$$

where N is the total number of words, and Del, Sub, and Ins are deletion, substitution, and insertion errors. Note that the word accuracy for domain-specific words is the primary criterion because they are more important for spoken dialogue systems than other words.

The ASR accuracies for the four models are listed in Table 4. We can see that accuracies of domain-specific words improved using the proposed method (LM #3) compared with the baselines (LMs #1 and #2). This is mainly because the occurrence probabilities of the domain-specific words were appropriately increased by the class N-gram model. This result shows that our approach is promising.

On the other hand, ASR accuracies for all words were almost equivalent, but slightly degraded from LM #1. This may be because inappropriate class labels caused some insertion errors. A comparison between LMs #3 and #4 implies that this slight degradation in the "All word" condition would improve if the accuracy of the class labeling was improved.

We also checked sentences for which ASR results became correct. As expected, our method was effective for obtaining not only the unknown domain-specific words but the context around the known domain-specific words. An example follows:

Correct:	*zyanru washoku wo sakujo shite kudasai*
	(Please delete the genre, Japanese-style.)
ASR (Model #1):	*zyanru wa choco wo sakujo shite kudasai*
ASR (Model #3):	*zyanru washoku wo sakujo shite kudasai*

Here, "*washoku* (Japanese-style)" is a domain-specific word in the GENRE class and is contained in the DB. The ASR result based on LM #1 failed because the context before and/or after "*washoku*" was not sufficiently trained. The ASR result based on LM #3 became correct because the context around GENRE was appropriately obtained.

5 Conclusion and Perspectives

In order to solve the issue where no sufficient training data are available when a new SDS is constructed, we prepared an initial training data set for the classifier by using simple string matching. We confirmed that unknown domain-specific words can be extracted from a similar-domain corpus by using the maximum entropy classifier. We then constructed a class N-gram LM by using the similar-domain corpus to which class information was given. The experimental results showed that our approach is promising; that is, the ASR accuracy for domain-specific words improved.

Two future tasks are planned. First, we need to improve the performance of the classifier described in Section 3. One possible solution is to formulate the problem as the sequence labeling and then to incorporate the conditional random field (CRF) [12]. The features used here also need to be investigated. Second, we will advance the "bootstrapping" cycle. The results reported in this paper are only after the first round. We need to confirm what will happen when more cycles are performed, and we also need to determine when to stop the cycle.

Acknowledgements. We would like to thank Yahoo Japan Corporation and the National Institute of Informatics (NII) for providing us "Yahoo! Chiebukuro data (2nd edition)." This research has been partly supported by the JST PRESTO Program.

References

1. Brown, P.F., deSouza, P.V., Mercer, R.L., Pietra, V.J.D., Lai, J.C.: Class-based n-gram models of natural language. Computational Linguistics 18(4), 467–479 (1992)
2. Sarikaya, R., Gravano, A., Gao, Y.: Rapid language model development using external resources for new spoken dialog domains. In: Proc. IEEE-ICASSP, pp. 573–576 (2005)
3. Sethy, A., Georgiou, P.G., Narayanan, S.S.: Building topic specific language models from webdata using competitive models. In: Proc. INTERSPEECH, pp. 1293–1296 (2005)
4. Misu, T., Kawahara, T.: A bootstrapping approach for developing language model of new spoken dialogue systems by selecting Web texts. In: Proc. INTERSPEECH, pp. 9–12 (2006)
5. Weilhammer, K., Stuttle, M.N., Young, S.: Bootstrapping Language Models for Dialogue Systems. In: Proc. INTERSPEECH, pp. 17–20 (2006)
6. Creutz, M., Virpioja, S., Kovaleva, A.: Web augmentation of language models for continuous speech recognition of SMS text messages. In: Proc. EACL, pp. 157–165 (2009)
7. Hakkani-Tur, D., Rahim, M.: Bootstrapping language models for spoken dialog systems from the world wide web. In: Proc. IEEE-ICASSP, vol. 1, pp. 1065–1068 (2006)
8. Varga, I., Ohtake, K., Torisawa, K., De Saeger, S., Misu, T., Matsuda, S., Kazama, J.: Similarity based language model construction for voice activated open-domain question answering. In: Proc. IJCNLP, pp. 536–544 (2011)
9. Berger, A.L., Pietra, V.J.D., Pietra, S.A.D.: A maximum entropy approach to natural language processing. Computational Linguistics 22(1), 39–71 (1996)
10. Kudo, T., Yamamoto, K., Matsumoto, Y.: Applying conditional random fields to Japanese morphological analysis. In: Proc. EMNLP, pp. 230–237 (2004)
11. McCallum, A.K.: Mallet: A machine learning for language toolkit (2002), http://mallet.cs.umass.edu/
12. Lafferty, J., McCallum, A., Pereira, F.: Conditional Random Fields: Probabilistic Models for Segmenting and Labeling Sequence Data. In: Proc. ICML, pp. 282–289 (2001)

A Speaker Diarization System with Robust Speaker Localization and Voice Activity Detection

Yangyang Huang, Takuma Otsuka, and Hiroshi G. Okuno

Abstract. In real-world auditory scene analysis of human-robot interactions, three types of information are essential and need to be extracted from the observation data – who speaks *when* and *where*. We present a speaker diarization system that is used to accomplish the resolution. Multiple signal classification (MUSIC) is a powerful method for voice activity detection (VAD) and direction of arrival (DOA) estimation. We propose our system and compare its performance in VAD and DOA with the method based on MUSIC algorithm.

1 Introduction

Robot auditory functions are expected to facilitate an intuitive and natural human-robot interaction such as for the situation shown in Fig. 1. In such a situation, the ability from observations to recognize who speaks when and where is necessary. We deal with this problem as a speaker diarization task. A situation of free speech among multiple speakers is considered, which means speech without any scenario.

Speaker diarization is essential for various applications. In [1], a 3D auditory scene visualizer is proposed. It shows meta information, such as speech text and direction of speakers. They are extracted from observed signals by constructing a speaker diarization system using an audition system for robots [2]. In [3], a speaker diarization method including DOA and automatic speech recognition (ASR) is presented to estimate automatically "who speaks when and what" for a group meeting situation.

The speaker diarization described in this paper differs from most systems mentioned in [4] by the presence of overlap speech in input audio signals. To deal with the overlap speech, DOA estimation, VAD, sound source separation and source identification methods have been developed [2, 3, 5].

Yangyang Huang · Takuma Otsuka · Hiroshi G. Okuno
Graduate School of Informatics, Kyoto University, Japan
e-mail: yangyang@kuis.kyoto-u.ac.jp

M. Ali et al. (Eds.): *Contemporary Challenges & Solutions in Applied AI*, SCI 489, pp. 77–82.
DOI: 10.1007/978-3-319-00651-2_11   © Springer International Publishing Switzerland 2013

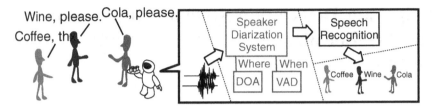

Fig. 1 Need for speaker diarization. Robot waiter needs to understand who orders when and where to deliver the drinks

We focus on DOA and VAD, aiming to detect *when* and *where* a speech segment occurs. In particular, as the processing is done by connecting a plurality of elementary technology items in series, the performance of the preceding stage will affect the subsequent processing. Using robust methods in the preceding processing is required for various observations. For example, in the robot audition system, HARK, a multiple signal classification (MUSIC) algorithm is used for DOA and VAD in preceding processing, and source separation processing is followed. However, parameters for DOA and VAD need to be selected carefully, as they affect the performance of the entire system.

We also evaluate the DOA and VAD for free speech in a real environment. Ground truth for speaker direction and active speech segments are necessary, therefore We have corrected the ground truth data by using the MAC3D system. In addition, we use the rate of precision and recall for VAD evaluation.

Aiming to obtain higher accuracy for the speaker diarization system, We have improved the overall performance by using the independent vector analysis (IVA), which is a robust way of separating sound sources.

2 Problem Statement and System Configuration

For an input of multi-channel audio signals and output of speech segments and speaker direction, we assume that the transfer function of the microphone is known. A transfer function represents the transfer characteristic of the sound from each direction to the microphone array.

The proposed processing flow is shown in Fig. 2. After a short-time Fourier transform of the multi-channel audio signal, we apply IVA for separating the observed signals. Then VAD by threshold processing is used on the separated voice, and DOA by using the MUSIC algorithm.

2.1 Blind Source Separation by IVA

The IVA method is an expansion of the blind source separation method, independent component analysis (ICA). This section will give an overview of ICA and then briefly describe the extension to IVA [6, 7].

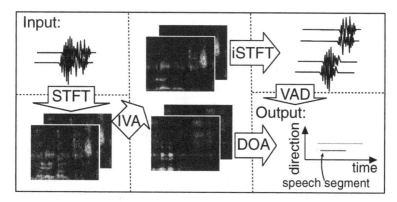

Fig. 2 Processing flow of two channel audio signal for two-speaker free speech. Lines in time-direction coordinates represent speech segments by two speakers.

In the ICA method, the observed signal in the time-frequency domain $\mathbf{Z}_{t,f} = [z_{t,f}^1, ..., z_{t,f}^M]^T$ is modeled as: $\mathbf{Z}_{t,f} = \mathbf{A}_f \mathbf{Y}_{t,f}$.

Here, $\mathbf{Y}_{t,f} = [Y_{t,f}^1, ..., Y_{t,f}^M]^T$ is the audio signal for sound sources in the tth time frame and fth frequency bin, $\mathbf{A}_f$ is a mixing matrix. For the signal $\{\mathbf{Z}_{t,f}\}_{t=1}^T$, we calculate a separation matrix $\mathbf{W}_f$ which satisfies that components of $\hat{\mathbf{Y}}_{t,f}$ are statistically independent. $\hat{\mathbf{Y}}_{t,f} = \mathbf{W}_f \mathbf{Z}_{t,f}$.

However, the ICA method has a permutation problem: $\hat{\mathbf{Y}}_{t,f}$ calculated for each frequency bin f may not in the same order as the original $\mathbf{Y}_{t,f}$. Thus, we need to select the correct component belonging to the same sound source in each frequency bin when restoring to the original audio signal.

By using this method, components of $\{\mathbf{Y}_{t,f}\}_{f=1}^F$ are deemed as an F-dimensional vector, this problem is avoided because $\{\mathbf{W}_f\}_{f=1}^F$ is optimized at the same time.

2.2 VAD by Threshold Processing

We detect speech segment in separated audio signals. First, we restore the separated time-frequency domain signal $Y_{t,f}$ to the time domain waveform signal y_t and extract one channel of them. Then we divide the waveform signal into shorter segment Δt with no overlap. For each segment, the number of samples with absolute power larger than T_v is calculated. A speech segment is confirmed if the number is more than T_s. The time-frequency domain signal corresponding to the determined speech segments are introduced into the next DOA process.

2.3 DOA by MUSIC Algorithm

This section explains the MUSIC algorithm in general and discusses the advantages of using it directly for separated audio signals. The output of MUSIC is the spectrum which contains the energy calculated for each direction θ in each block b.

For time-frequency domain signals, the correlation matrix is calculated and averaged in a time block as: $\mathbf{R}_{b,f} = \sum_{t=(b-1)*\Delta T}^{b\Delta T} \mathbf{z}_{t,f}\mathbf{z}_{t,f}^H$, where H denotes a conjugate transpose operator. The eigenvalue decomposition of $R_{b,f}$ is given by $\mathbf{R}_{b,f} = \mathbf{E}_{b,f}\Lambda_{b,f}\mathbf{E}_{b,f}^{-1}$, where $\mathbf{E}_{b,f}$ is the eigenvector matrix and $\Lambda_{b,f}$ is the eigenvalue matrix. The eigenvector and eigenvalue can also be represented as $\lambda_{b,f,m}$ and $\mathbf{e}_{b,f,m}$.

The MUSIC spectrum is calculated as: $P_{b,f,\theta} = \frac{\|\mathbf{a}_{f,\theta}\mathbf{a}_{f,\theta}^H\|}{\sum_{m=N+1}^{M}|\mathbf{a}_{f,\theta}\mathbf{e}_{b,f,m}^H\|}$ where $a_{f,\theta}$ is the transfer function for the θth direction and fth frequency bin, and N denotes the number of sound source. A detailed explanation of MUSIC algorithm is provided in [8].

In the HARK system, Threshold processing is applied to the MUSIC spectrum calculated. The number of sound sources parameter and threshold parameter is used.

The use of IVA in our proposed system avoids the request for the parameter of the number source and threshold for the MUSIC spectrum.

3 Experiments

This section consists of three parts:

(1) Collection of reference data and create ground truth for free speech
(2) Evaluation criteria for VAD and DOA result
(3) Comparison of the baseline method and proposed method

3.1 Reference Data and Ground Truth

The ground truth for DOA and VAD takes a two-dimensional-array form represented as $x_{b,\theta}$. The indexes of the array denote the time block bins b and the direction bins θ. The value of x denotes speaker ID or 0 for silence segment.

We obtained data of four free speech records and its reference data for evaluation. Each record was 240 seconds. The sampling rate was 16000 Hz, the block length for VAD was 0.5s, and the length of the STFT was 512 points.

3.2 Evaluation Criteria

DOA and VAD result is evaluated in frame wides. Impressively, how estimated result $\hat{x}_{b,\theta}$ is close to the ground truth $x_{b,\theta}$. Let the number of speech segments in ground truth by system be S_a, the number of speech segments in ground truth be S_d, the number of speech segments detected correctly be S_c, which is the difference between the correctly detected speech segment and the corresponding speech segment in ground truth. Evaluation criteria are defined as follows:

Precision: $R_p = \frac{S_c}{S_a}$, Recall: $R_r = \frac{S_c}{S_d}$, F measure: $F = \frac{2R_pR_r}{R_p+R_r}$.

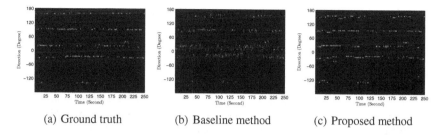

| (a) Ground truth | (b) Baseline method | (c) Proposed method |

Fig. 3 Comparison of DOA and VAD result

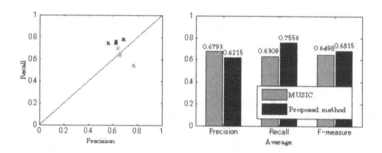

Fig. 4 Quantitative result. Recall rate improved by over 10 percent. Green points in left figure show result by baseline method. Blue points shows result obtained by using the proposed method, with $T_v = 0.01$ and $T_s = 100$

3.3 Results and Comparison

In the IVA source separation process, the source number is 5, the number of speakers, and this number remains unchanged during the process. Threshold T_v and T_s for VAD is set to 0.01 (in 0~1) and 100 (in 0~8000).

Detection result for active speech segments are shown in Fig. 3. A comparison of the two methods is shown in Fig. 3. The proposed method has a higher F measure result and recall rate. Although precision rate is lower, clearly it can be improved by integrating other sensors.

4 Conclusion

We proposed a speaker diarization system which includes various processes connected in series. By using the IVA separation method as in the preceding process, the proposed system has the following advantages: (1) Higher recall and F measure than those of the baseline method. For a speaker diarization system, recall rate is the main requirement, which is equal to reduce the deletion error. (2) With the unchanged parameters, the proposed system still performs better than the baseline method in which parameters need to be fixed to improve performance.

Acknowledgements. This study was partially supported by a Grant-in-Aid for Scientific Research (S) 24220006.

References

1. Kubota, Y., Yoshida, M., Komatani, K., Ogata, T., Okuno, H.G.: Design and implementation of 3d auditory scene visualizer towards auditory awareness with face tracking. In: Tenth IEEE International Symposium on Multimedia, pp. 468–476 (2008)
2. Nakadai, K., Takahashi, T., Okuno, H.G., Nakajima, H., Hasegawa, Y., Tsujino, H.: Design and implementation of robot audition system 'hark' open source software for listening to three simultaneous speakers. Advanced Robotics 24(5-6), 739–761 (2010)
3. Araki, S., Hori, T., Fujimoto, M., Watanabe, S., Yoshioka, T., Nakatani, T., Nakamura, A.: Online meeting recognizer with multichannel speaker diarization. In: ASILOMAR, pp. 1697–1701 (2010)
4. Tranter, S.E., Reynolds, D.A.: An overview of automatic speaker diarization systems. Proceedings of the IEEE Transactions on Audio, Speech, and Language Processing 14(5), 1557–1565 (2006)
5. Nakamura, K., Nakadai, K., Asano, F., Ince, G.: Intelligent sound source localization and its application to multimodal human tracking. In: Proceedings of the IEEE/RSJ International Conference on IROS, pp. 143–148 (2011)
6. Hyvarinen, A., Karhunen, J., Oja, E.: Independent Component Analysis. Wiley Interscience (2001)
7. Ono, N.: Stable and fast update rules for independent vector analysis based on auxiliary function technique. In: IEEE Workshop on Applications of Signal Processing to Audio and Acoustics, pp. 189–192 (2011)
8. Schmidt, R.: Multiple emitter location and signal parameter estimation. IEEE Transactions on Antennas and Propagation 34(3), 276–280 (1986)

A Content Fusion System Based on User Participation Degree on Microblog

Wo-Chen Liu[1], Meng-Hsuan Fu[1], Kuan-Rong Lee[2], and Yau-Hwang Kuo[1,3]

Abstract. Microblog users generally publish their opinions by using condensed text with some non-textual content. Besides, post responses from participants often include noise such as chaotic messages or unrelated information to the theme. Thus, we propose a Feature-based Filtering Model attempts to filter these noises. Moreover, we propose a method, which select the responses based on user participation degree, Maximum Discussion Group Detection (MDGD), to solve the problem of ignored information by current content fusion approaches. Briefly, the posts with higher user participation degree are selected to extract the short text from original post and its responses. The related content from several microblog platforms is also referred to enrich the fusion results. In the experiments, the test data set is collected from the microblog platforms of Plurk and Facebook. Finally, the Normalized Discounted Cumulative Gain (NDCG) metrics show that our method is capable to provide qualified extraction results.

Keywords: Microblog, Short Text, Content Fusion.

1 Introduction

Rapid growth of microblog services usages has induced numerous research topics such as opinion sentiment analysis, commercial advertisement, summarization and

Wo-Chen Liu · Meng-Hsuan Fu · Yau-Hwang Kuo
Department of Computer Science and Information Engineering,
National Cheng Kung University, Tainan, Taiwan
e-mail: {alex58237332,mhfu,kuoyh}@ismp.csie.ncku.edu.tw

Kuan-Rong Lee
Department of Information Engineering, Kuan Shan University, Tainan, Taiwan
e-mail: leekr@mail.ksu.edu.tw

Yau-Hwang Kuo
Department of Computer Science, National Chengchi University, Taipei, Taiwan

M. Ali et al. (Eds.): *Contemporary Challenges & Solutions in Applied AI*, SCI 489, pp. 83–90.
DOI: 10.1007/978-3-319-00651-2_12   © Springer International Publishing Switzerland 2013

so on. Automatic Text Summarization (ATS) addresses both the problem of finding the most important portion of document and generating coherent summary result from document. The existing ATS techniques may not appropriate for microblog write-ups fusions [1] because those adopted traditional Natural Language Processing techniques, which only cope with the pure textual information. In addition, existing content fusion approaches ignore the relationships between the responses and the parent post that will increase the problem of information deficiency [2]. In microblog posts, the textual expression is limited. Removing non-textual elements will lose extensive useful information.

In order to help users to grasp content with hot topics from different social media services in a short period. Firstly, a Behavior-based Feature Extraction Model to extract the contextual and non-contextual features from the posts and their responses are presented. Then, the Feature-based Filtering Model is proposed to filter the useless content and reserve the useful content. Next, the Maximum Discussion Group Detection (MDGD) Model is designed to find the highly discussion group from the dataset. Besides, a multiple source selection model called Multiple Posts Selection Model to enrich the fusion content is also proposed. Finally, the Multi-source Fusion Model provides qualified fused results based on user query.

This paper is organized as follows. In next section, some background and related works are surveyed. The model of Multi-Feature Analysis for Microblog Content Fusion is described in Section 3, and the experimental results are shown in Section 4. In last section, the conclusion and future work of this paper are given.

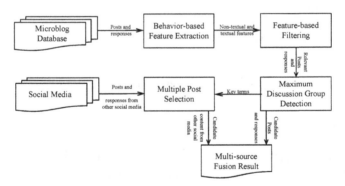

Fig. 1 Overall structure of Multi-feature Analysis for Microblog Content Fusion

2   Background and Related Works

The task of document summarization is to create a summary of many documents in order to provide the key points to readers. Some ATS systems are using semantic analysis, natural language processing technique or comparing the similarities between sentences by n-gram or entropy [1,2,3].

Summarization is divided into abstraction and extraction: An abstraction contains at least some sentences that do not exist in the original document. An extraction is text spans (typically, paragraphs or sentences) selected from the original document. Abstract approaches are more difficult than extraction because it has to change the sentence position or create a new sentence for result, not only picks up the important sentence. Abstract approaches must apply many NLP techniques to deal with the sentences such as segmentation, tagging, parsing, nouns filtering, syntax tree construction and so on. Extract approaches are usually using statistical model or learning method to extraction sentence rather than using natural language processing techniques, such as Term Frequency and Inverse Document Frequency (TF-IDF), entropy calculation, relevant analysis, topic extraction and so on.

3 Multi-feature Analysis for Microblog Content Fusion

3.1 Behavior-Based Feature Extraction Model

The past microblog researches mainly focus on textual content analysis. However, non-textual content usually contain more useful information in microblog posts. Thus, the non-textual features are also taken into account in this stage. According to the composition of the microblog property, the extracted features that we referred are as follows:

- The features about a target post including the number of emoticons, pictures, videos, hyperlinks, responses, likes, reposts and the length of this post.
- The features about each response to the target post including the delta time between this response and the target post, the number of emoticons, pictures, videos, hyperlinks, and the length of this response.

3.2 Feature-Based Filtering Model

We observe that responses often contain some recurring or irrelevant content. These responses are usually spread by the specific users that are robots or advertisers. Therefore, this model attempts to classify each post into two categories, which are Irrelevant Content and Relevant Content. According to the composition of the microblog property, we select the features to train classifier.

3.3 Maximum Discussion Group Detection Model

We observe that the existing posts do not always have the most responses in microblog. According to the composition of user behavior property, the concept of Discussion Graph (DG) about the responses to a post is proposed. For a given post and its responses, $U = \{u_1, u_2, ..., u_n\}$ denotes as a set of N users in responses.

$G = (V, E)$ denotes as an indirect graph, where node V is corresponding to the N users, edge E indicates the "mention relationships" between these nodes, and the number of node V is denoted as $|v|$. The weight of edge E is corresponding to the total number of discussion series, and it adds one when the discussions (mention relationships) occurred in a post, which denotes as e_{weight}. The DG is performed to find maximum connected component in post i called Maximum Discussion Group (MDG), which denotes as MDG_i. If the number of $MDG_i > 2$ in post i, fetching the group which has maximal sum of weights; randomly fetching one group if the maximal sum of weights are the same. The objective of MDG is used to rank these posts according to the participation degree with the given query.

Equation (1) is used to quantify the participation degree for a post from the MDG_i. Participation Score (PS_i) is a quantitative function for counting the participation rate of a post. Combining with PS_i, equation (2) considers the number of responses, which denotes as $pres_i$ and the number of likes, which denotes as $plike_i$ in the post i. The basic idea of the Information Score (IS_i) is performed to find the "highly discussed post"; the higher IS_i causes by the higher user participation degree. Finally, we fetch top β results of input post and reserve the highly frequency terms, which weight $tfw_k = \dfrac{|c_k| * 100}{\sum_{\forall c_j \in C} |c_j|} > \gamma$ from MDG.

$$PS_i = \begin{cases} \dfrac{\sum_{v \in MDG_i} e_{weight}^2}{(|v|_{\forall v \in MDG_i}) - 1}, & \text{if} |v|_{\forall v \in MDG_i} \geq 2 \\ 0, & \text{if} |v|_{\forall v \in MDG_i} < 2 \end{cases} \tag{1}$$

$$IS_i = \propto * plike_i + (1 - \propto) * pres_i * PS_i \tag{2}$$

3.4   Multiple Post Selection Model

This model enriches the content by exploiting other platforms. Firstly, we obtain the key terms from previous steps. Secondly, we apply Behavior-based Feature Extraction Model, Feature-based Filtering Model and MDGD model query by key terms to eliminate noise and measure the user participation degree between posts. Finally, fetch the highest participation degree post as candidate content from each platform.

3.5   Multi-source Fusion Model

In this section, we gather information and then extract representative features of a post and its responses by score. A candidate post p_i and its responses R_i, which has many representative words same as post p_i, denote as $PR = \{pr_i | pr_i = p_i \cup R_i, i \in ID\}$.

Thus, after the sentence segmentation, we calculate score for each sentence. Equation (5) is presented the score formula of sentence s_j, where $StScore(s_j)$ is the sentence score of the terms entailed by s_j. Note that, all of the terms k must consist in both sentence s_j and its parent post p_i .Equation (3) is the time-weighting function and equation (4) is the length-weighting function.

$$TW(j) = 1 - \log_{|S|} j \tag{3}$$

$$LW(|s_j|) = e^{\frac{(|s_j|-\mu)^2}{2\sigma^2}} \tag{4}$$

$$StScore(s_j) = \frac{\sum_{tfw_k \in s_j \wedge tfw_k \in p_i} tfw_k}{LW(|s_j|)} * TW(j) \tag{5}$$

When the score $StScore(s_j)$ is calculated, we fetch the top ψ sentences. If the length of fusion results is smaller than φ, union candidate sentences from Multiple Post Selection model. Finally, the output is the fusion results of the original post and its responses.

4 Experimental Results

4.1 Feature-Based Filtering

In order to evaluate our model, we gathered a set of testing data by collecting posts and responses via Plurk official API from 2011 February to 2011 March. The 10-fold cross-validation based on SVM with RBF kernel was used, the nine randomly generated folds in each round were used for module construction and the one fold was reserved for testing. Macro-Averaging and Micro-Averaging recognition rate were used to evaluate overall performance. The recognition rate was 96.8% in the Relevant Content and 88.7% in the Irrelevant Content. As shown in figure 2, the micro-average recognition rate was 95.3% and macro-average one was 92.7%. Based on the learned weights of features, we observed that the mentioned feature is not very useful in determining Relevant of Irrelevant Content.

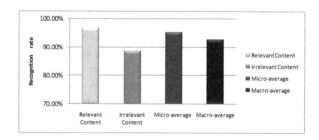

Fig. 2 Performance of the Feature-based Filtering model

4.2 Maximum Discussion Group Detection

For the experiments, we used the Normalized Discounted Cumulative Gain (NDCG) metric [4]. NDCG was adopted to measure the effectiveness of our method. Each post was judged on the scale 0-2, where 0 denotes "irrelevant", 2 denotes "completely relevant", and 1 means "somewhere in-between".

In experiment, we used Plurk API to acquire 29,208 posts and 153,163 responses on 2011 December and analyzed the sensitivity of the parameter in our system. The best performance achieved when $\alpha = 0.5$, $\beta = \lfloor |input\ post| * 0.1 \rfloor$, $\mu = 35$, $\sigma = 88.983$, and $\gamma > 2$. The results were shown in Table 1. The NDCG score was obtained by the top ten results of input post. From the results, "highly discussed post" was obtained by *IS*, which can quantify the participation degree of a post from MDG. The method we proposed achieves higher objective and accurate in the experimental results.

Table 1 Measurement of user satisfying performance for the MDGD model

Query Terms	NDCG@10$_{IS}$	NDCG@10$_{pres}$
蔡英文 (the one 2012 Taiwanese president candidate)	0.92	0.81
馬英九 (the other 2012 Taiwanese president candidate)	0.93	0.51
林書豪 (Jeremy Shu-How Lin, an basketball player in NBA)	0.71	0.52

4.3 Multi-source Fusion

Currently, there are no similar benchmarks or measurement of comparisons for microblog fusion system. In order to conduct the user satisfactions, we recruited 4 Social Network experts and 16 Plurk users. Same as previous steps, we evaluated the performance of multi-source fusion results including the same query terms from Plurk and Facebook. The user satisfactions were shown in figure 3, we used

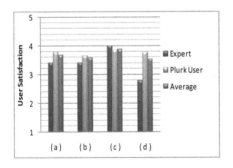

(a) I understand the main topic of the posts from the fusion result before read the original posts and its responses.

(b) I feel the fusion results are complete after read the original content.

(c) I am able to follow the fusion results easily.

(d) I feel the fusion results are useful and clear.

Fig. 3 Overall measurements of user satisfaction performance

5-point rating scale (5 = strongly agree, 1 = strongly disagree) for each question. For question (a), we obtained 3.7 point, 3.6 point for question (b), 3.9 point for question (c) and 3.55 point for question (d) in average. The results shown that our system is capable for providing good quality of fused results based on user query.

5   Conclusions

This paper addressed the issue of automatically selecting passages from microblog posts and provides fusion results. These problems are difficult because (i) responses are not always discussing the same topic, and (ii) content length is limited and often including noise in microblog posts.

In this paper, we utilize the "mention relationship" to rank the posts according to the participation degree with the given query. According to the ranking of discovered MDGs, we fused multiple posts and investigated how to find the candidate posts correctly via contextual features and non-contextual features. Finally, we evaluated our system on Plurk and Facebook, and observed how users' satisfaction could correlates with the effectiveness of fusion result by NDCG.

Presently, the textual content of microblogs often includes Network Informal Language (NIL) [5],[6]. In the future, the contextual features will be mainly focus on. We will experiment on MMSEG [7] to support NIL terms. Unlike informal language, many new NIL are created continually. New terms detection is one of the most important issues for social information processing. Moreover, experiments will explore the optimal combination of various features such as "social relations", "user opinion" and so on to enhance the user satisfactions.

Acknowledgments. This work was supported in part by the National Science Council of Taiwan under grant NSC-100-2221-E-006-251-MY3 and Southern Taiwan Science Park Administration under grant 101CC03.

References

1. Weng, J.-Y., Yang, C.-L., Chen, B.-N., Wang, Y.-K., Lin, S.-D.: IMASS: an intelligent microblog analysis and summarization system. In: Proceedings of the 49th Annual Meeting of the Association for Computational Linguistics: Human Language Technologies: Systems Demonstrations (HLT 2011), pp. 133–138. Association for Computational Linguistics, Stroudsburg (2011)
2. Harabagiu, S., Hickl, A.: Relevance Modeling for Microblog Summarization. In: Proceedings of the Fifth International AAAI Conference on Weblogs and Social Media (ICWSM 2011), Barcelona, Spain (2011)
3. Grogory Silber, H., McCoy, K.F.: Efficiently computed lexical chains as an intermediate representation for automatic text summarization. Comput. Linguist. 28(4), 487–496 (2002), doi:10.1162/089120102762671954

4. Järvelin, K., Kekäläinen, J.: Cumulated gain-based evaluation of IR techniques. ACM Trans. Inf. Syst. 20(4), 422–446 (2002), doi:10.1145/582415.582418
5. Yun-Qing, X., Kam-Fai, W., Wei, G.: NIL is not Nothing: Recognition of Chinese Network Informal Language Expressions. In: Proceedings of the Fourth SIGHAN Workshop at International Joint Conference on Natural Language (IJCNL 2005), pp. 95–102 (2005)
6. Zhang, X., Yao, T.: A Study of Network Informal Language Using Minimal Supervision Approach. In: Autonomous Systems – Self-Organization, Management, and Control, pp. 169–175 (2008)
7. MMSEG: A Word Identification System for Mandarin Chinese Text Based on Two Variants of the Maximum Matching Algorithm, http://technology.chtsai.org/mmseg/

Network Intrusion Detection System Based on Incremental Support Vector Machine

Haiyi Zhang, Yang Yi, and Jiansheng Wu

Abstract. Based on simple incremental SVM, we proposed an improved incremental SVM algorithm (ISVM), and combined it into a kernel function U-RBF and applied it into network intrusion detection. The simulation results show that the improved kernel function U-RBF has played some role in saving training time and test time. The ISVM has eased the oscillation phenomenon in the process of the learning to some extent, and the stability of ISVM is relatively good.

1 Introduction

Intrusion detection, a proactive real-time security protection technology, has got more and more attention. It is a network security technology that is used to detect acts, which damage or attempt to damage the confidentiality, integrity or availability of the system or network. It can effectively make up the shortages of firewalls, data encryption, authentication and other static defense strategies and can carry out a full range of network security protection by combining with other network security products as well.

Research on the incremental SVM learning algorithm in the intrusion detection system is of significance. Firstly, it is difficult for traditional security and defense strategies to meet the ever-changing needs of network security and short of static protection technologies. Secondly, current intrusion detection methods are mostly non-incremental learning algorithm. As the accumulation of the new incremental samples, the training time expenses will continue to increase. Thirdly, incremental learning can rapidly learn from the new incremental samples to modify the existing model. Time consumption is relatively small. Finally, compared with non-incremental algorithms, incremental learning algorithms are in a relatively small number of studies, especially the incremental SVM algorithm.

Haiyi Zhang
Jodrey School of Computer Science, Acadia University, Wolfville, Noca Scotia,
Canada B4P 2R6
e-mail: haiyi.zhang@acadiau.ca

Yang Yi · Jiansheng Wu
Sun Yat-sen University, China

M. Ali et al. (Eds.): *Contemporary Challenges & Solutions in Applied AI*, SCI 489, pp. 91–96.
DOI: 10.1007/978-3-319-00651-2_13   © Springer International Publishing Switzerland 2013

Due to the oscillation phenomenon that simple incremental SVM will lead to in the follow-up learning process, we propose an improved incremental SVM algorithm ISVM. In order to reduce the number of samples and shorten the training time, the algorithm introduces sample selection process, and applies it to network intrusion detection. Compared to other algorithms, the test results indicate that the ISVM algorithm eases the oscillation phenomenon to some extent in the incremental learning process, and also saves training and prediction time.

2   Incremental Support Vector Machine

Support vector machines are a new type of learning machine based on statistical learning theory and structural risk minimization principle. The basic idea is to make a nonlinear mapping from the input space to high dimensional space. Then construct a classification hyper plane that separates the training data by a maximal margin.

The standard SVM requires that all samples be trained at the same time. If new samples are added, the SVM needs to be retrained to find a new optimal classification hyper plane, so the tolerance to noise for the SVM is not high. On the other hand, it is very difficult to obtain a complete training set in the beginning if lack of initial samples. The accuracy of the learning machine will be affected, so we hope the learning machine can continuously improve the learning accuracy by using the priori knowledge when new samples are added in.

Based on the above issues, there are a number of researchers working on the SVM with incremental learning function. Syed et al [1,4] first proposed the incremental learning algorithm of SVM. It divided the training set into N subsets, after training on a subset, the algorithm only retained the support vectors and discarded the other samples, and then added them into next subset to form a new training subset and trained on the new training subset.

3   Incremental Support Vector Machine Based on Reserved Set

3.1   Kernel Function U-RBF (Unitizing RBF)

We propose a new kernel function, U-RBF, according to the consideration unitizing the record. The U-RBF adds the mean value of the feature attributes and the mean square deviation to the RBF, which will cut down the training time.

3.2   The KKT Conditions and the Distribution of the Samples

In order to obtain the decision function when training the SVM, we have to solve the quadratic programming problem, where we need to use the optimality conditions of the optimization problem, Karush-Kuhn-Tucker (KKT) conditions.

The solving of SVM boils down to the solving of the convex quadratic programming problem, where the KKT conditions are defined as:

$$y_i f(x_i) - 1 \begin{cases} \geq 0 & \alpha_i = 0 \\ = 0 & 0 < \alpha_i < C \\ \leq 0 & \alpha_i = C \end{cases} \tag{1}$$

The KKT conditions (1) are equivalent to (2):

$$\begin{cases} \alpha_i = 0 \Rightarrow f(x_i) \geq 1 & or \quad f(x_i) \leq -1 \\ 0 < \alpha_i < C \Rightarrow f(x_i) = 1 & or \quad f(x_i) = -1 \\ \alpha_i = C \Rightarrow -1 \leq f(x_i) \leq 1 \end{cases} \tag{2}$$

We can get the classifier from the process of training the samples, where α_i is the Lagrange multiplier that the sample corresponds to. According to (2), we know that the samples that meet the requirement of the KKT conditions are the support vectors.

3.3 Reserved Set Strategy

In the simple incremental SVM, when the new incremental samples come, first, check whether all the samples meet the KKT conditions. The follow-up learning will lead to oscillation phenomenon if the initial samples are insufficient. This paper presents a reserved set strategy that will retain. In order to ease the oscillation phenomenon in the follow-up learning process, select samples from the reserved set to combine them with those samples that are contrary to the KKT conditions in the new incremental sample set and the original support vector set to form a new training set, according to the weight. After incremental learning, update the reserved set and the weights of the samples.

There are two key issues: one is how to select samples to construct the reserved set; the other is how to empower the value for each sample.

4 Simulated Experiments

4.1 Experiments Description

We take the benchmark KDD-CUP99 with nearly 5 million network records, as the dataset of the experiments. Each record is a grouped sequence that starts and terminates within the required timeframe when in line with the stated protocols. It also has a class identifier, which denotes either normal class or some specific attack class. There are 22 attack classes divided into the four categories.

In this paper, the detection rate, false alarm rate and correlation coefficient are used as the evaluation indicators for the intrusion detection. The purpose of the incremental SVM proposed in this paper is not only to enhance the intrusion detection rate and reduce false alarm rate, but also to reduce the training time and the forecast time as much as possible. So, the training time and the forecast time are adopted as the evaluation indicators as well.

The simulation experiments are divided into two parts. In the first part we mainly verify the effectiveness of the improved kernel function U-RBF by comparing the effectiveness of U-RBF with RBF and POLY. I in the second part we mainly test the detection performance of the RS-ISVM by comparing it with the simple incremental SVM and the peer-to-peer incremental SVM.

4.2   Experiments of RS-ISVM

Our experiments are performed to verify the effectiveness of the improved SVM. First, the dataset is randomly divided into two subsets, each contains both normal and abnormal class, one is the source of the training data, and the other is the source of test data. Secondly, select 9 data sets at randomly, named I1 to I9, from the training subset as the incremental training set randomly, each set contains 300 normal samples and 300 abnormal samples, and any two training samples sets are not mixed. Thirdly, select the normal records and attack records, with the number of normal records are equal to the attack records, as the test set.

Table 1 The performance of the RS-ISVM using different parameters

Parameters		Training set (cc)		
η_1	η_2	I_1 (cc)	I_2 (cc)	I_3 (cc)
0.1	10	0.767	0.759	0.768
0.1	20	0.767	0.798	0.814
0.2	10	0.767	0.765	0.769
0.2	20	0.767	0.799	0.815
0.3	10	0.767	0.766	0.771
0.3	**20**	**0.767**	**0.806**	**0.821**
0.4	10	0.767	0.749	0.737
0.4	20	0.767	0.029	0.785
0.5	10	0.767	0.784	0.766
0.5	20	0.767	0.103	0.753

Because the strategy of selecting samples involves the scale factors η_1 and η_2, which need to be set, we choose the best combination of η_1 and η_2 by assessing algorithm performance with combinations of different parameters. Different scale factors η_1 and η_2 have been chosen for simulation and the comparison results of correlation coefficients are listed in Table 1, where $I_1 \sim I_3$ are the incremental training subsets and cc denotes the correlation coefficient.

The best combination of the parameters η_1 and η_2 has been marked in bold. First, train the SVM on the sample set I_1, and then the incremental SVM will be updated when the new incremental sample sets $I_2 \sim I_9$ arrive. This paper will compare the RS-ISVM with the simple incremental SVM (Simple-ISVM) [2,5] and the peer-to-peer incremental SVM (KKT-ISVM) [3,6] with the detection rate, false alarm rat, correlation coefficient, training time and forecast time. Since all the three methods are based on the C-SVM, all the three algorithms will use the RBF as the kernel functions. Moreover, RS-ISVM also uses U-RBF.

Table 2 Comparison of the training time and test time

	RS-ISVM(U)		RS-ISVM		Simple-ISVM		KKT-ISVM	
	TrD(s)	TeD(s)	TrD(s)	TeD(s)	TrD(s)	TeD(s)	TrD(s)	TeD(s)
I1	0.438	1.469	1.094	4.656	1.016	5.64	1.105	5.688
I2	0.718	1.985	2.078	6.218	3.844	14.813	7.547	16.203
I3	1.015	2.719	3.156	9.266	5.625	20.562	16.0	23.984
I4	1.328	3.391	5.797	11.719	9.093	23.922	30.437	29.656
I5	1.719	5.546	6.781	13.641	18.86	26.656	38.297	34.297
I6	0.984	1.00	8.156	15.532	18.047	35.047	48.688	38.547
I7	0.438	1.105	8.609	16.938	22.672	42.094	71.64	44.563
I8	0.453	1.11	10.485	18.969	33.094	47.969	83.203	50.016
I9	0.553	1.11	13.812	21.328	28.14	43.813	96.266	53.047

The comparison results of the training time and test time are listed in Table 2, where TrD and TeD represent the training time and test time respectively. As shown in Table 2, KKT-ISVM needs the most training time, because it has to do cross judging and more training. The training time of Simple-ISVM and RS-ISVM is in the acceptable range. However, RS-ISVM(U) has obvious advantages in training time, because it is clearly much shorter than RS-ISVM.The above results show that the improved kernel function U-RBF plays some role in saving the training and test time. On the other hand, compared to the Simple-ISVM and KKT-ISVM, the changes that RS-ISVM do to the original classifier are more reliable, and it will not cause large fluctuations in detection performance to the classifier. Moreover, with cumulative incremental training on the new sample set, RS-ISVM continuously improves the detection performance.

5   Conclusions

We have proposed a strategy that is based on the reserved set, which retains those non-support vectors that are most likely to become support vectors to ease the oscillation phenomenon in the process of the incremental learning. At the same time, a concentric circle method has been proposed to select the samples to construct the reserved set. Lastly, an incremental SVM algorithm RS-ISVM that is based on the reserved set bas been proposed.

References

[1] Wang, X.D., Zheng, C.Y., Wu, C.M., Zhang, H.D.: New algorithm for SVM-based incremental learning. Computer Applications 10(26), 2440–2443 (2006)
[2] Laskov, P., Gehl, C., Kruger, S., Muller, K.: Incremental support vector learning: Analysis, Implementation and application. Journal of Machine Learning Research 7, 1909–1936 (2006)
[3] Shilton, A., Palamiswami, M., Ralph, D., Tsoi, A.: Incremental training of support vector machines. IEEE Transactions on Neural Networks 16, 114–131 (2005)
[4] Cheng, S., Shih, F.: An improved incremental training algorithm for support vector machines using active query. Pattern Recognition 40, 964–971 (2007)
[5] Liang, Z.Z., Li, Y.F.: Incremental support vector machine learning in the primal and applications. Neurocomputing (February 20, 2009)
[6] Deng, N.Y., Tian, Y.J.: A new method of data mining – support vector machines. Science Press (2004)

Use of Fuzzy Information for Heterogeneous Performance Evaluation

Mohammad Anisseh[*] and Mohammad Reza Shahraki

Abstract. Personnel performance appraisals have been practiced in many organizations and institutions with the purpose for salary adjustments, promotions, training, and other decisions that affect employee status in the company. Human judgments, including preferences are often vague and cannot be estimated in exact numerical values. This paper uses a method under the linguistic framework for heterogeneous performance evaluation, which allocates different weights for assessor members to use linguistic terms in order to express their fuzzy preferences for candidate solutions and for individual judgments. The introduced method has been used in the empirical study, and the results have been analyzed.

Keywords: Performance evaluation, Group decision making, Fuzzy numbers.

1 Introduction

Fan and Zhang [4] and Chuu [5] stated that human beings are faced with issues of decision making that basically involves choosing the most-preferred alternatives from a limited set of alternatives to obtain certain-predefined objectives. Evaluating personnel is one of the most critical decisions that must be taken [6]. Most crucial and significant decisions in organizations are made by groups of managers or experts. There are two types of group decision making: (a) homogeneous and (b) heterogeneous. Contrary to homogeneous group decision making, the

Mohammad Anisseh
Department of Industrial Management, Imam Khomeini International University, Iran
e-mail: manisseh@ikiu.ac.ir

Mohammad Reza Shahraki
Department of Industrial Engineering, University of Sistan and Baluchestan, Iran
e-mail: reza_mohammad87@yahoo.com

[*] Corresponding author.

M. Ali et al. (Eds.): *Contemporary Challenges & Solutions in Applied AI*, SCI 489, pp. 97–103.
DOI: 10.1007/978-3-319-00651-2_14   © Springer International Publishing Switzerland 2013

heterogeneous decision making considers opinions from decision makers that constitutes of different gender, age, education, functional specialization and expertise [7-9]. In such an environment, disagreement always happens in group decision making as members in a group generally do not come to the same decision [10]. To solve disagreements for one decision maker that involves multi criteria evaluation and ranking problems, the multi criteria decision making (MCDM) has been developed [11-16]. Crisp data are insufficient to simulate real life situations and managers' judgments normally include preferences are often vague and not precise estimates of the numerical value [17]. Chuu [5] stated that the traditional MCDM methods are random processes and deterministic, and unable to solve group decision making problems with inaccurate and vague information. Therefore, fuzzy MCDM methods were developed. The concept of fuzzy sets is one of the most important and significant instruments in the computational intelligence [18]. Fuzzy sets and fuzzy logic are powerful mathematical tools for modelling in uncertain systems in industry, nature and humanity and act as facilitators for common-sense reasoning in decision making in the absence of complete and accurate information [19]. Therefore, designing applicable multi-dimensional appraisal systems for heterogeneous appraisers has been a main concern for scholars. This paper is trying to use heterogeneous group decision making model under fuzzy environment for personnel performance appraisal.

2 Preliminaries

Definition 2.1. *A* fuzzy set presents a boundary with a gradual contour, by contrast with classical sets, which present a discrete border. Let U be the universe of discourse and u a generic element of U, then $U = \{u\}$. A fuzzy subset Ã, defined in U, is: $\tilde{A} = \{(u, \mu_{\tilde{A}}(u)) \mid u \in U\}$, Where $\mu_{\tilde{A}}(u)$ is designated as membership function or membership grade of u in Ã[20].

Definition 2.2. A is a fuzzy number, if A is normal and convex [21]. A triangular fuzzy numbers can be expressed as M = (l, m, u), where $l \leq m \leq u$, in which $l \leq m \leq u$. In the fuzzy event, parameters (l, m, u) are the smallest, promising, and the largest possible value, correspondingly [22]. Equation (1) describes the triangular fuzzy number membership function M, when $l=m=u$, it is a non-fuzzy number by agreement as shown in Fig. 1 [23]

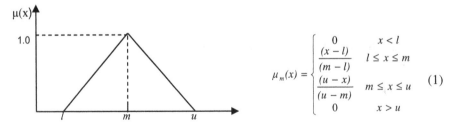

$$\mu_m(x) = \begin{cases} 0 & x < l \\ \dfrac{(x - l)}{(m - l)} & l \leq x \leq m \\ \dfrac{(u - x)}{(u - m)} & m \leq x \leq u \\ 0 & x > u \end{cases} \quad (1)$$

Fig. 1 The triangular fuzzy membership function [2]

Definition 2.3. For two triangular fuzzy numbers M_1 and M_2 of the main operational laws are as follows [24]:

$$M_1 + M_2 = (l_1 + l_2, m_1 + m_2, u_1 + u_2), \qquad \omega \times M_1 = (\omega l_1, \omega m_1, \omega u_1), \ \omega > 0, \ \omega \epsilon R,$$
$$M_1 \times M_2 = (l_1 \times l_2, m_1 \times m_2, u_1 \times u_2), \qquad M_1^{-1} = (1/u_1, 1/m_1, 1/l_1). \tag{2}$$

Definition 2.4. A variable with values of word or sentences in an artificial language is defined a linguistic variable [25]. After the cardinality of the linguistic terms set are recognized, linguistic terms and semantics that must be arranged as show in (Tables 1 & 2)..

Table 1 Linguistic variables for the ratings [1]

Very Poor	VP	(0, 0, 1)
Poor	P	(0, 1, 3)
Medium Poor	MP	(1, 3, 5)
Fair	F	(3, 5, 7)
Medium Good	MG	(5, 7, 9)
Good	G	(7, 9, 10)
Very Good	VG	(9, 10, 10)

Table 2 Linguistic variables for the importance weight of each criterion [3]

Very Low	VL	(0, 0, 0.1)
Low	L	(0, 0.1, 0.3)
Medium Low	ML	(0.1, 0.3, 0.5)
Medium	M	(0.3, 0.5, 0.7)
Medium High	MH	(0.5, 0.7, 0.9)
High	H	(0.7, 0.9, 1.0)
Very High	VH	(0.9, 1.0, 1.0)

3 The Fuzzy Heterogeneous Performance Evaluation Method

The purpose of this method is to enhance group agreement on the group decision making outcome based on Borda count. Let $A = \{A_1, A_2, \ldots, A_m\}$ be a discrete set of alternatives, $P = \{P_1, P_2, \ldots, P_k\}$ be the set of decision makers, and $\lambda = (\lambda_1, \lambda_2, \ldots, \lambda_p)$ be the weight vector of decision makers, where $\lambda_p \geq 0$, $P = 1, 2, \ldots, k$, and $\sum_{p=1}^{k} \lambda_p = 1$. Let $C = \{C_1, C_2, \ldots, C_n\}$ be the set of attributes, and $w = (w_1, w_2, \ldots, w_n)$ be the weight vector of attributes, where $w_n \geq 0$, $n = 1, 2, \ldots, j$, $\sum_{n=1}^{j} w_n = 1$ [26]. The fuzzy group decision problem can be concisely expressed as matrix format [27]:

$$\tilde{P}_t = \begin{matrix} & C_1 & C_2 & \cdots & C_n \\ A_1 & \tilde{x}_{11} & \tilde{x}_{12} & \cdots & \tilde{x}_{1n} \\ A_2 & \tilde{x}_{21} & \tilde{x}_{22} & \cdots & \tilde{x}_{2n} \\ \vdots & \vdots & \vdots & \vdots & \vdots \\ A_m & \tilde{x}_{m1} & \tilde{x}_{m2} & \cdots & \tilde{x}_{mn} \end{matrix} \tag{3}$$

$\tilde{W} = [\tilde{w}_1, \tilde{w}_2, \cdots, \tilde{w}_n]$ Where $\tilde{x}_{ij}^k$ and $\tilde{w}_j^k$ are linguistic variables that can be shown by fuzzy numbers as shown in (Tables 1, 2).

The proposed models are linearly described in the following 11 steps:

1- Identifying evaluation criteria. 2- Generating alternatives. 3- Identifying weights of criteria and weights of decision makers. 4- Presenting preferences on the part of each decision maker (every decision maker gives preferences to per alternative based on every attribute according to linguistic terms such as Table 2.

5- Construction of fuzzy decision matrix. In fuzzy decision matrix, we suppose that, each $\tilde{x}_{ij}^k$ is fuzzy number. 6- Construct the normalized fuzzy decision matrix that can be found in [28, 29]. If $(\tilde{x}_{ij}, i = 1,2,\dots,m, j = 1,2,\dots,n)$ are triangular fuzzy numbers, then the normalization process can be performed by [25]:

$$\tilde{r}_{ij} = \left(\frac{a_{ij}}{c_j^*}, \frac{b_{ij}}{c_j^*}, \frac{c_{ij}}{c_j^*}\right) \quad \tilde{r}_{ij} = \left(\frac{a_j^-}{c_{ij}}, \frac{a_j^-}{b_{ij}}, \frac{a_j^-}{a_{ij}}\right) \quad i = 1,2,\dots,m, \quad j \in C \quad (4)$$

Where B and C are the set of benefit criteria and cost criteria, respectively. 7- Construction of defuzzification decision matrix; the defuzzified value of fuzzy number can be obtained from Equation

$$BNP_i=[(U_i - L_i)+(M_i - L_i)]/3+L_i , \forall i \quad (5)$$

8- Considering proper value (DM weights) of every decision making group member idea $N_{ij\lambda} = N_{ij} \times \lambda_p$ (6). N_{ij} is an element of defuzzification decision matrix for every DM, and λ_p is the weight of per DM idea. 9- Formation of R_j matrixes; while the rows of the matrix are alternatives and its columns are DMs opinions based on j criterion. So n matrixes in lieu of j attributes were established (R_j):

$$R_j = \begin{matrix} A_1 \\ \vdots \\ A_i \\ \vdots \\ A_m \end{matrix} \begin{bmatrix} r_{1,j}^1 & \cdots & r_{1,j}^p & \cdots & r_{1,j}^k \\ \vdots & & \vdots & & \vdots \\ r_{i,j}^1 & \cdots & r_{i,j}^p & \cdots & r_{i,j}^k \\ \vdots & & \vdots & & \vdots \\ r_{m,j}^1 & \cdots & r_{m,j}^p & \cdots & r_{m,j}^k \end{bmatrix} \begin{matrix} , i=1,2,\dots,m \\ , j=1,2,\dots,n \\ , p=1,2,\dots,k \end{matrix} \quad (7)$$

Computing linear sum in lieu of P decision makers $\left(\sum_{p=1}^{k} r_{i,j}^p\right)$ and final grade of every alternative in lieu of j attributes would be calculated.

In this matrix the line with the highest mark is the first rank and the line with the lowest mark is m rank.

$$R_G = \begin{matrix} & C_1 & \cdots & C_j & \cdots & C_n \\ A_1 \\ \vdots \\ A_i \\ \vdots \\ A_m \end{matrix} \begin{bmatrix} r'_{1,1} & \cdots & r'_{1,j} & \cdots & r'_{1,n} \\ \vdots & & \vdots & & \vdots \\ r'_{i,1} & \cdots & r'_{i,j} & \cdots & r'_{i,n} \\ \vdots & & \vdots & & \vdots \\ r'_{m,1} & \cdots & r'_{m,j} & \cdots & r'_{m,n} \end{bmatrix} \quad (8)$$

10- Changing R_G matrix into Borda count, i.e. alternative with first rank based on per criterion would have $m-1$ relative value on the basis of m alternatives. The same goes for, alternative with second rank ($m-2$ relative value). Alternatives with m rank would receive zero relative values. We multiply the Borda count matrix with the corresponding weight vector of attributes function [2]

$$\begin{matrix} A_1 \\ A_2 \\ \vdots \\ A_m \end{matrix} \begin{bmatrix} b_{1,1} & b_{1,2} & \cdots & b_{1,n} \\ b_{2,1} & b_{2,2} & \cdots & b_{2,n} \\ \vdots & \vdots & \ddots & \vdots \\ b_{m,1} & b_{m,2} & \cdots & b_{m,n} \end{bmatrix} \times \begin{bmatrix} w_1 \\ w_2 \\ \vdots \\ w_n \end{bmatrix} = \begin{bmatrix} b_{1,1}.w_1 + & b_{1,2}.w_2 + & \cdots & b_{1,n}.w_n \\ b_{2,1}.w_1 + & b_{2,2}.w_2 + & \cdots & b_{2,n}.w_n \\ \vdots & \vdots & \ddots & \vdots \\ b_{m,1}.w_1 + & b_{m,2}.w_2 & \cdots & b_{m,n}.w_n \end{bmatrix} = \begin{bmatrix} r_1 \\ r_2 \\ \vdots \\ r_m \end{bmatrix} \quad (9)$$

The alternative sum with the highest value would be considered as the first rank and the lowest represents the last rank.

4   Empirical Study

Table 3 Fuzzy performance measurement of assessor group done by professionals

Profes-sional Team	Assessors			
	Manager	Colleagues	Inferior	Employee him/herself
kjkjlProf_1	VH	MH	MH	H
Prof_2	H	MH	M	ML
Prof_3	VH	H	MH	H
Prof_4	VH	H	VH	H
Prof_5	H	M	MH	L
Prof_6	H	M	M	MH
Prof_1	0.9,1.0,1.0	0.5,0.7,0.9	0.5,0.7,0.9	0.7,0.9,1.0
Prof_2	0.7,0.9,1.0	0.5,0.7,0.9	0.3,0.5,0.7	0.1,0.3,0.5
Prof_3	0.9,1.0,1.0	0.7,0.9,1.0	0.5,0.7,0.9	0.7,0.9,1.0
Prof_4	0.9,1.0,1.0	0.7,0.9,1.0	0.9,1.0,1.0	0.7,0.9,1.0
Prof_5	0.7,0.9,1.0	0.3,0.5,0.7	0.5,0.7,0.9	0.0,0.1,0.3
Prof_6	0.7,0.9,1.0	0.3,0.5,0.7	0.3,0.5,0.7	0.5,0.7,0.9
Total	5.500	4.133	4.067	3.733
Weight	0.32	0.24	0.23	0.21

This study was performed in Energy Efficiency Organization active in the field of energy consumption optimizing. Twenty-one persons were selected as a professional group according to the president of EEO organization opinion based on their experiences and education in order to determine the criteria, weights, and assessors' viewpoint weights as shown in Table. 3. In this paper the organization's middle managers from deputy of training and optimizing energy consumption were evaluated. So each manager was assessed by 4 assessors based on 33 quality and quantity attributes. A_1, A_2, A_3, A_4 was allocated to 4 managers. In step 4 of algorithm method, the assessors presented preferences for each alternative based on each attribute according to linguistic terms based on Table. 4. Next step constructed the normalized fuzzy decision matrix and converted the normalized fuzzy decision matrix to the defuzzification decision matrix by (Equation (6) as shown in (Table. 5). Step 8, 9: Considering proper value (DM weights) of every decision making group member idea by (Equation (7) and establishing n matrixes lieu of j attribute as (Table. 6). Step 10: In (Table. 6) Linear sum would be reached in lieu of P decision makers and final rank of every alternative in lieu of j attribute would be calculated. In these matrixes, the line with the highest mark is the first rank and the line with the lowest mark is m rank. Step 11: We change the R_G matrix into Borda count; multiply the Borda count matrix with the corresponding weight vector of attributes by (Equation (9)). The alternative sum with the highest value would be considered as the first rank and the lowest represents the last rank. The ordinal ranks of four alternatives (Middle managers) are attained as follows: $A_3 \gg A_4 \gg A_1 \gg A_2$. Therefore, A_3 is the optimal candidate.

Table 4 The ratings of four candidates by decision makers under all criteria

	C_1				C_2				C_3			
	A_1	A_2	A_3	A_4	A_1	A_2	A_3	A_4	A_1	A_2	A_3	A_4
P_1	MG	G	F	G	G	G	F	VG	G	G	VG	VG
P_2	G	G	G	VG	MG	G	MG	VG	VG	VG	G	VG
P_3	F	F	G	MG	F	F	G	G	G	G	VG	G
P_4	G	G	VG	MG	MG	G	VG	G	VG	VG	VG	VG

Table 5 The fuzzy normalized decision matrix and criteria weights

		C_1		C_2		C_3	
P_1	A_1	0.5,0.7,0.9	0.700	0.7,0.9,1.0	0.867	0.7,0.9,1.0	0.867
	A_2	0.7,0.9,1.0	0.867	0.7,0.9,1.0	0.867	0.7,0.9,1.0	0.867
	A_3	0.3,0.5,0.7	0.500	0.3,0.5,0.7	0.500	0.9,1.0,1.0	0.967
	A_4	0.7,0.9,1.0	0.867	0.9,1.0,1.0	0.967	0.9,1.0,1.0	0.967
P_2	A_1	0.7,0.9,1.0	0.867	0.5,0.7,0.9	0.700	0.9,1.0,1.0	0.967
	A_2	0.7,0.9,1.0	0.867	0.7,0.9,1.0	0.867	0.9,1.0,1.0	0.967
	A_3	0.7,0.9,1.0	0.867	0.5,0.7,0.9	0.700	0.7,0.9,1.0	0.867
	A_4	0.9,1.0,1.0	0.967	0.9,1.0,1.0	0.967	0.9,1.0,1.0	0.967
P_3	A_1	0.3,0.5,0.7	0.500	0.3,0.5,0.7	0.500	0.7,0.9,1.0	0.867
	A_2	0.3,0.5,0.7	0.500	0.3,0.5,0.7	0.500	0.7,0.9,1.0	0.867
	A_3	0.7,0.9,1.0	0.867	0.7,0.9,1.0	0.867	0.9,1.0,1.0	0.967
	A_4	0.5,0.7,0.9	0.700	0.7,0.9,1.0	0.867	0.7,0.9,1.0	0.867
P_4	A_1	0.7,0.9,1.0	0.867	0.5,0.7,0.9	0.700	0.9,1.0,1.0	0.967
	A_2	0.7,0.9,1.0	0.867	0.7,0.9,1.0	0.867	0.9,1.0,1.0	0.967
	A_3	0.9,1.0,1.0	0.967	0.9,1.0,1.0	0.967	0.9,1.0,1.0	0.967
	A_4	0.5,0.7,0.9	0.700	0.7,0.9,1.0	0.867	0.9,1.0,1.0	0.967
Criteria weights		0.0335		0.0335		0.0299	

Table 6 Aggregation matrix (R_G) based on per criterion (Middle managers)

C_1	P_1	P_2	P_3	P_4	Σ	R	C_2	P_1	P_2	P_3	P_4	Σ	R
A_1	0.22	0.20	0.11	0.18	0.72	4	A_1	0.27	0.16	0.11	0.14	0.70	4
A_2	0.27	0.20	0.11	0.18	0.78	2	A_2	0.27	0.20	0.11	0.18	0.78	2
A_3	0.16	0.20	0.19	0.20	0.77	3	A_3	0.16	0.16	0.19	0.20	0.73	3
A_4	0.27	0.23	0.16	0.14	0.81	1	A_4	0.30	0.23	0.19	0.18	0.92	1

C_{33}	P_1	P_2	P_3	P_4	Σ	R
A_1	0.27	0.23	0.19	0.20	0.91	3,4
A_2	0.27	0.23	0.19	0.20	0.91	3,4
A_3	0.30	0.20	0.22	0.20	0.94	2
A_4	0.30	0.23	0.19	0.20	0.94	1

$$
\begin{array}{c}
A_1 \\
A_2 \\
A_3 \\
A_4
\end{array}
\begin{bmatrix}
4 & 4 & \cdots & 3,4 \\
2 & 2 & \cdots & 3,4 \\
3 & 3 & \cdots & 2 \\
1 & 1 & \cdots & 1
\end{bmatrix}
\Rightarrow
\begin{bmatrix}
0 & 0 & \cdots & 0.5 \\
2 & 2 & \cdots & 0.5 \\
1 & 1 & \cdots & 2 \\
3 & 3 & \cdots & 3
\end{bmatrix}
\times
\begin{bmatrix}
0.0335 \\
0.0335 \\
\vdots \\
0.0299
\end{bmatrix}
=
\begin{bmatrix}
1.259 \\
0.663 \\
2.055 \\
2.023
\end{bmatrix}
=
\begin{bmatrix}
3 \\
4 \\
1 \\
2
\end{bmatrix}
$$

Table 7 Comparison of proposed method and pervious personnel performance appraisal (Middle managers)

Candidates	A_1	A_2	A_3	A_4
Previous assessment grade	29.5	29	29.5	29.5
Fuzzy heterogeneous performance evaluation method	3	4	1	2

5 Conclusion

The personnel evaluation is one of the most important and complicated aspects of human resource management. A new proposed personnel performance appraisal model was used in this study, in which personnel are evaluated from different points of view and evaluation's errors are minimized. In multi criteria group decision making with linguistic variables, the assessors may have vague information, limited attention and different information processing capabilities. This paper uses a fuzzy group decision making method which allows group members to express their fuzzy preferences in linguistic terms for candidate selection and for individual judgments. The proposed method covered heterogeneous performance evaluation by considering the decision makers' viewpoint weights. The results of the mentioned models are compared with the pervious personnel performance appraisal model (the evaluation of the subordinate from manager point of view) Table . 7.

References

[1] Wade, T.: Optimum Dielectric Selection Using a Weighted Evaluation Factor. Semicond. Int. 18 (1995)
[2] Anisseh, M., Piri, F., Shahraki, M.R., et al.: Fuzzy Extension of TOPSIS Model for Group Decision Making Under Multiple Criteria. Artificial Intelligence Review (2012)
[3] Xu, Z.S., Chen, J.: An interactive method for fuzzy multiple attribute group decision making. Information Sciences 177(1), 248–263 (2007)

Part VI
Optimization

Designing Loss-Aware Fitness Function for GA-Based Algorithmic Trading

Yuya Arai*, Ryohei Orihara, Hiroyuki Nakagawa,
Yasuyuki Tahara, and Akihiko Ohsuga

Abstract. In these days, an algorithmic trading in stock or foreign exchange (henceforth forex) market is in fashion, and needs for automatically performing stable asset management are growing. Machine learning techniques are increasingly used to construct trading rules of the algorithmic trading, as researches on the algorithmic trading advance. Our study aims to build an automatic trading agent, and in this paper, we concentrate in designing a module which determines trading rules by machine learning. We use Genetic Algorithm (henceforth GA), and we build trading rules by learning parameters of technical indices. Our contribution in this paper is that we propose new fitness functions in GA, in order to make them robuster to change of market trends. Although profits were used as a fitness function in the previous study, we propose the fitness functions which pay more attention to not making a loss than to gaining profits. As a result of our experiment using real TSE(Tokyo Stock Exchange) data for eight years, the proposed method has outperformed the previous method in terms of gained profits.

1 Introduction

In these days, financial transactions, such as stock trading and forex, are performed on web. Rule-based automatic or algorithmic trading is becoming popular. Researches to create trading rules by means of machine learning techniques are being carried out.

However, since the financial market is complicated, the newest application of data mining and the machine learning techniques are not progressing much. Our study aims to build trading rules using the machine learning technique, and evaluation is performed with real TSE(Tokyo Stock Exchange) data for eight years.

Yuya Arai · Ryohei Orihara · Hiroyuki Nakagawa · Yasuyuki Tahara · Akihiko Ohsuga
Graduate School of Information Systems, The University of Electro-Communications,
Tokyo, Japan
e-mail: arai-y@ohsuga.is.uec.ac.jp

* He is currently working at Fuji Electric IT Solutions Co., Ltd.

M. Ali et al. (Eds.): *Contemporary Challenges & Solutions in Applied AI*, SCI 489, pp. 107–114.
DOI: 10.1007/978-3-319-00651-2_15   © Springer International Publishing Switzerland 2013

Researches on the prediction for a stock have been done widely. In the researches for stock price prediction, ones which use ANN [1] [2], fuzzy theory [3] [4] [5], and so on have been done, and the result is obtained. There are several kinds of prediction, for example, the prediction of stock closing price of the next day, and that of a long-term trend. As for inputs to a prediction algorithm, values such as the past stock closing price and a technical index are used.

Researches on an algorithmic trading have been done, although there are not many. In the research using Genetic Network Programming [6], a stock closing price and technical indices are input, and trading rules are built. In the research using Genetic Algorithm [7] [8], trading rules are built by evaluating the parameters. Moreover, the framework which performs automatic trading of a stock called Kaburobo [9] is offered. The contest is done there which evaluates the trading algorithm from various viewpoints.

Our contributions are as follows: We found out the problem of overfitting in the previous work. In order to avoid the problem, we proposed fitness functions which favor small loss rather than large profits. By the evaluation experiment using real TSE 136 stocks data, profits of the proposed method doubled compared with the previous method.

This paper is organized as follows: Section 2 introduces the previous work[7] on algorithmic trading. In Section 3 the proposed method which devised the loss-aware fitness function is explained. The method is evaluated through the experiment using the past stock price data. In Section 4 the contents of this paper are summarized.

2 Previous Work

2.1 The Method and Its Performance

In this paper, an automatic trading method using GA [7] is used as a start point. The flow of the method and its performance are explained below.

Their automatic trading system is built using four technical indices. As shown in Table 1, a technical index is calculated with the past stock prices; it expresses the state of the price of one stock. Since there are various kinds of technical indices, it is difficult to determine the technical index used for the constructing of the system in general. Each index has a few parameters like Table 1, and it is not easy to determine them too. In our approach, the parameters of the indices are determined by GA, and our study aims to build an automatic trading system with technical analysis using the optimal parameters.

As a result of the experiment [7], the profits were negative on the first year of 6-year evaluation period, but they improved from the second year on, and were positive in the end. However, upon comparing optimized parameters resulted in positive profits in test data, and the tendency of overfitting was observed. The profits in the evaluation period were only 38% of those in the training period. Therefore, designing to prevent the overfitting at the time of training was mentioned as a future subject.

Table 1 Technical indices

Indices	Parameters
Simple Moving Average(SMA)	short period, long period
Exponential Moving Average(EMA)	short period, long period
Bollinger Bands(BB)	period, factor
Channel Break Out(CBO)	period

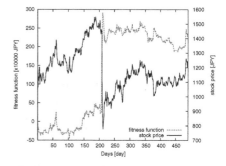

Fig. 1 The change of the fitness function from October, 2001 to September, 2003

2.2 Discussion and Possible Improvement

In the previous work, the fitness function was defined as profits, and the optimized parameters resulted in positive profits in test data. Although GA avoids a local optimum more effectively than other machine learning algorithms, it cannot be completely free from overfitting.

Figure 1 shows two time series, which are the fitness function in one gene and a stock price, where the fitness function is profits, during two years. The stock price plunges approximately on the 200th day and the fitness function rises suddenly at the same time. However, the fitness function gradually decreases in other periods. The trading rule using this gene works only in a particular condition. The system therefore may not be able to perform well to a general stock price movement.

When the fitness function is profits, the overfitting could result where the function adapts to particular cases in which large profits can be gained. Such a function is inadequate to yield a universally usable trading rule. Thus, an appropriate fitness function is the one which, unlike the fitness function shown in Figure 1, is hardly affected even if there is a rapid change of a stock price. In this paper, the fitness functions are created by the method described in Section 3.1.

3 Proposed Method

3.1 Loss-Aware Fitness Functions

In order to avoid overfitting, the following fitness functions, and aspects of loss, such as the winning percentage of trading and the preference for small loss over large profits, are considered: Here fit0 is used in the previous method, and fit1 to fit 3 are used in the proposed method.

fit 0 $Profits$

fit 1 $Profits \times WinningPercentage$

fit 2 $Profits - Loss$

fit 3 $-Loss + 0.01 \times Profits$

Where

- *Profits* is defined as the accumulation of *Gain*. *Gain* is defined as the amount which results after subtracting the amount of acquisition and a commission from the amount of sale.
- *Winning Percentage* is defined as the ratio of trading opportunities where the amount of sale is greater than the amount of acquisition.
- *Loss* is defined as an absolute value of the accumulated *Gain*, which considers only cases where *Gain* is negative.

In fit1, by compensating *Profits* with *Winning Percentage*, the smaller the number of times of failed trading becomes, the larger the fitness function becomes. In fit2., *Loss* is subtracted from *Profits*, therefore it puts the twice importance to *Loss* compared with *Profits*. Consequently, the smaller the loss becomes, the larger the fitness function becomes. In fit3., it is important for the system not to result in the loss. *Profits* will be used in order to break a tie, when *Loss* is the same amount. The experiments are performed using the three fitness functions with *Profits* as baseline, and the difference in effectiveness will be compared between the proposed method and the previous method.

3.2 Condition of Experiment

The effectiveness of the proposed method explained in Section 3.1 is evaluated by comparing it with the previous method. Here, an experimental condition is explained.

The experiment is done using the stock price data for eight years from 2001 to 2008. The parameters are adjusted to the optimal combination by applying GA to the 2-year training data. The system sells and buys using the trained trading rule, for the subsequent one month. The system is evaluated as following. The period of the first two years and the subsequent one month are counted as one window, the experiment is performed on the total period of 72 windows, sliding the window one month at a time.

An initial fund is assumed to be infinite, because portfolio issue is not treated in this paper. Amount of trade is considered the minimum trading unit for each stock, and a commission is assumed to be 1000 JPY per trading. Four technical indices explained in Table 1 are used. The term for each technical index takes a value from the set of $\{5, 10, 15, 20, 25, 30, 50, 75, 100, 200\}$, and the factor for Bollinger Bands also takes the value from a set of $\{1.0, 1.5, 2.0, 2.5, 3.0\}$. Moreover, out of 225 stocks quoted as Nikkei Average, 136 stocks which met the following conditions for the full period are used as stock price data.

- The stock has continued to be listed in Nikkei Average.
- There has been no change of the trading unit.
- There has been no share splitting.

The conditions of GA are as follows. The number of individuals is set to 100 and the generation number is set to 500. Minimal Generation Gap [10] model is used as the generation alternation model. Uniform crossover [11], and uniform mutation [12] are employed.

Four fitness functions explained in Section 3.1 are evaluated. The stock price data of the Yahoo Finance [13] are used for the experiment.

3.3 Result

Table 2 and Figure 2 are the results of the experiment for the four fitness functions and average profit over 136 stocks. With fit0, which is based only on profits, the profits were small in the end despite it produced some gains during the period. On the other hand, as for fit1 and fit2, which applied compensation to the fitness function, although the loss came out in the first half of the period, they bounced back and finished trading for six years in almost identical success. As for fit3, which attaches great importance to not taking a loss, it made relatively small loss during the period when the other fitness functions took large loss, therefore it obtained the largest profits in the end. Profits using fit3 doubled compared with the previous method, fit0.

3.4 Discussion

Figure 3 shows time series of a stock price and fitness functions as in Figure 1. Figure 3 shows time series of the fitness functions fit1, fit2 and fit3 as well as one of fit0. Fit0, fit1 and fit2 reacted superfluously to a stock price approximately on the

Table 2 Profit average of 136 stocks every fitness function and technical index[x10,000 JPY]

	SMA	EMA	BB	CBO	fit ave
fit 0	16.9	17.0	-3.1	-2.8	7.0
fit 1	18.7	25.8	5.2	2.0	12.9
fit 2	19.5	19.9	3.0	8.8	12.8
fit 3	22.2	25.4	6.7	8.7	15.8
tec ave	19.3	22.0	2.9	4.2	

Table 3 Genes of buying

indices	Parameters(fit0)	Parameters(fit3)
SMA	15 day, 30 day	15 day, 50 day
EMA	10 day, 15 day	15 day, 50 day
BB	100 day, 2.0	75 day, 3.0
CBO	5 day	50 day

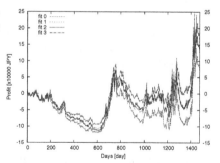

Fig. 2 The result of the experiment. The time series of profit average of 136 stocks. The vertical axis expresses profit, and the cross axis expresses time(six years).

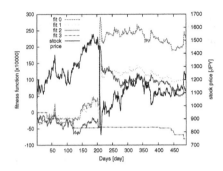
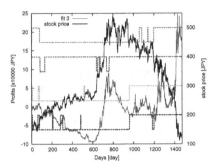

Fig. 3 The change of the fitness function from October, 2001 to September, 2003

Fig. 4 The profits and dealings the four technical indices

200th day, when the stock price plunged. On the other hand, fit3, which is designed not to take out a loss, showed no reaction. This is an example where fit3 is not overfitted toward an extreme price movement. This could explain why the good result was obtained.

Moreover, a large difference appeared also in the frequency of transactions. The number of transactions in the training period of Figure 3 was 262 times in fit0; on the other hand, it was 34 times in fit3. It is thus thought that, in order not to take out a loss with fit3, the trading rule which lessens the frequency of transactions was learned.

Genes of buying for fit0 and fit3 are compared in Table 3. In fit0, the terms for SMA, EMA and CBO are shorter than their counterpart in fit3. It could have yielded frequent changes from a *buy* decision to a *sell* and vice versa, triggerred relatively reactive trading decisions and resulted in the frequent tradings in fit0. As shown here, the difference in the parameters yielded the difference in trading timings and final profits.

Figure 4 shows the profits with fit3 and a stock price, along with buy / sell decisions shown in blue, purple, green and pink lines, based on the four technical indices, SMA, EMA, BB, and CBO respectively. The lines oscillate between two values. When they are in the lower value, the system decides to buy. Otherwise it decides to sell.

During the first half of the 6-year evaluation period, the stock price has been generally in an upward trend, particularly true between the second and the third year. The system has learned the trend and decided to buy most of the time, except BB. In the second half, especially the final year, the stock price has plunged. Despite of this the system has been able to produce profits because it has quickly learned the downward trend. It is this adaptability why the system has yielded the satisfactory results in both upward and downward trends.

4 Conclusion

In this paper, an improved method to design a fitness function for the algorithmic trading using GA has been presented. As shown in Section 3.3, the fitness function, which gave the top priority to not issuing a loss, has yielded good results, and there was a large improvement, compared with the previous study. Especially, profits of fit3 doubled compared with fit0.

There are some works to be done. First, the fitness function should be further improved. Although our method has achieved the certain improvement, and the period used for the experiment contains reasonably various stock market conditions including *Lehman Crash*, we cannot prove that it is universally optimum over various patterns of price movements. We need to have an effective design guideline which can be used for various time series trend.

Second, applications to the other financial markets, such as forex, should be considered. There might be necessary to take each market's characteristics into account to build an algorithm.

Acknowledgements. The authors would like to express their deepest gratitude to associate professor Hiroaki Oku of the University of Electro-Telecommunications who provided helpful comments and suggestions.

References

1. Lu, C.-J., Chiu, C.-C., Yang, J.-L.: Integrating nonlinear independent component analysis and neural network in stock price prediction. In: Chien, B.-C., Hong, T.-P., Chen, S.-M., Ali, M. (eds.) IEA/AIE 2009. LNCS, vol. 5579, pp. 614–623. Springer, Heidelberg (2009)
2. Neto, M.C.A., Calvalcanti, G.D.C., Ren, T.I.: Financial time series prediction using exogenous series and combined neural networks. In: IJCNN 2009, pp. 2578–2585 (2009)
3. Yu, T.H.K., Huarng, K.H.: A Bivariate Fuzzy Time Series Model to Forecast the TAIEX. Expert Systems with Applications 34, 2945–2952 (2008)
4. Chen, C.D., Chen, S.M.: A New Method to Forecast the TAIEX Based on Fuzzy Time Series. In: Proceedings of the 2009 IEEE International Conference on Systems, Man, and Cybernetics, San Antonio, Texas, pp. 3550–3555 (2009)
5. Chen, S.-M., Chu, H.-P.: TAIEX forecasting based on fuzzy time series and the automatically generated weights of defuzzified forecasted fuzzy variations of multiple-factors. In: Pan, J.-S., Chen, S.-M., Nguyen, N.T. (eds.) ICCCI 2010, Part II. LNCS, vol. 6422, pp. 441–450. Springer, Heidelberg (2010)
6. Chen, Y., et al.: A portfolio optimization model using Genetic Network Programming with control nodes. Expert Syst., 10735–10745 (2009)
7. Matsui, K., Sato, H.: A comparison of genotype representations to acquire stock trading strategy using genetic algorithms. In: ICAIS 2009, pp. 129–134 (2009)
8. Matsui, K., Sato, H.: Neighborhood evaluation in acquiring stock trading strategy using genetic algorithms. In: SoCPaR 2010, pp. 369–372 (2010)
9. KABU ROBO, http://www.kaburobo.jp/

10. Sato, H., et al.: A New Generation Alternation Model of Genetic Algorithms and Its Assessment. J. of Japanese Society for Artificial Intelligence 12(5), 734–744 (1997)
11. Syswerda, G.: Uniform Crossover in Genetic Algorithms. In: Proceedings of the Third International Conference on Genetic Algorithms, pp. 2–9 (1989)
12. Goldberg, D.E.: Genetic Algorithms in Search, Optimization, and Machine Learning. Addison-Wesley, Reading (1989)
13. Yahoo!finance, http://finance.yahoo.co.jp/

Watching Subgraphs to Improve Efficiency in Maximum Clique Search

Pablo San Segundo, Cristobal Tapia, and Alvaro Lopez

Abstract. This paper describes a new technique referred to as *watched subgraphs* which improves the performance of BBMC, a leading state of the art exact maximum clique solver (MCP). It is based on watched literals employed by modern SAT solvers for Boolean constraint propagation. The paper proposes to watch two subgraphs of critical sets during MCP search to efficiently compute new steps and bounds. Reported results validate the approach as the size and density of problem instances rise, while achieving comparable performance in the general case.

1 Introduction

A *clique* (alias complete graph) is a simple graph with all its vertices pairwise adjacent. The problem of knowing whether a clique subgraph of k vertices exists in a given graph is an NP-complete problem known as *k-clique*. The corresponding optimization problem is the *maximum clique problem* (MCP) whose goal is finding the largest possible clique. MCP is known to have important practical applications related to matching and has been related to many fields such as combinatorial chemistry, computer vision, global robot localization [1] etc.

In recent years research on MCP has produced a number of very efficient exact algorithms which have improved average performance in more than two orders of magnitude [2-6]. Of these, current leading algorithms are MCS [4] and bit parallel BBMC [5-6]. Specifically, Prosser in a comparison paper [7] reports BBMC as fastest for a number of instances in well known benchmarks. A release version of BBMC for Win-64bit S.O. is publicly available [8].

This paper describes a new idea which improves BBMC performance in large dense graphs while introducing minor overhead in the general case. We refer to

Pablo San Segundo
Universidad Politécnica de Madrid (UPM), Centro de Automática y Robótica (CAR-CSIC)
e-mail: pablo.sansegundo@upm.es

Cristobal Tapia · Alvaro Lopez
Universidad Politécnica de Madrid, Spain

M. Ali et al. (Eds.): *Contemporary Challenges & Solutions in Applied AI*, SCI 489, pp. 115–122.
DOI: 10.1007/978-3-319-00651-2_16   © Springer International Publishing Switzerland 2013

the new idea as *watched subgraphs*. The idea is based in *watched literals* used in modern SAT solvers for boolean constraint propagation as in [9]. In MCP, watched subgraphs help to improve empty set detection, as well as vertex selection and critical bit mask computations at each step.

The remaining part of the paper is structured as follows: sections 2 and 3 deal with definitions and related work in the field; section 4 presents the new algorithm BBMCW which uses watched sets; section 5 reports empirical results and finally section 6 shows conclusions and summarizes contribution.

2   Preliminaries and Notation

A simple undirected graph $G = (V, E)$ consists of a finite set of vertices V and edges E containing pairs of distinct vertices ($E \subseteq V x V$). $N(v)$ refers to the neighbor set of v, i.e. the set of adjacent vertices. For a subset of vertices U, $G(U)$ denotes the subgraph induced by U. Additional standard notation used in the paper for graph invariants include $\deg(v)$ for vertex degree (i.e. $|N(v)|$), $\Delta(G)$ for graph degree (i.e. $\max_{v \in V}(\deg(v))$) and $\omega(G)$ for the size of the maximum clique in G.

Vertex coloring is another classical problem in graph theory; its goal is to achieve a (minimum) label assignment (usually referred to as *coloring*) $c(v) : V \to \mathbb{N}$ such that adjacent vertices must all have different labels. Any graph coloring of size n (an *n-coloring*) $C(G) = \{C_1, C_2, \cdots, C_n\}$ partitions V in n disjoint *color classes* C_i (i.e. $v \in C_i \Leftrightarrow c(v) = i$). For a given vertex subset $U \subseteq V$, $C(U)$ or $C(G(U))$ will denote a vertex coloring of U.

Vertex coloring is very much related to efficient MCP search because the size of any feasible coloring $|C(G)|$ is an upper bound on the size of any maximum clique in G, i.e.:

$$|C(G)| \geq \omega(G) \tag{1}$$

3   Reference Algorithm and Related Work

Procedure REF in Table 1 describes the outline of recent efficient MCP algorithms and corresponds roughly to MCQ [2]. Variables used in REF include:

- U: The remaining subgraph hanging from the current node. When a subscript is added (i.e. U_v) it refers to the new remaining child subgraph after expanding v.
- $C(U)$ or C when it is clear from the context: A vertex coloring of U.
- S: The currently growing clique (path of the search).
- S_{max}: The maximum clique found at any moment.

REF enumerates all possible maximal cliques, one per branch. It starts with empty sets S and S_{max}. At each node a new candidate vertex is selected in step 2 and added to S (step 5) until a leaf node is found, when S is a maximal clique in G and is stored in S_{max}. Each time a new leaf node is reached, S is evaluated to see if it can unseat the current champion in S_{max} (step 7). On backtracking, REF deletes v from S (step 10) and the whole process is repeated until all possible maximal cliques have been enumerated.

U refers to the set of candidate vertices at each node, i.e. those that may enlarge the current clique S in a given step. The new candidate vertex is picked from U (step 2) and the new child subgraph $G(U_v)$ is computed (step 6) by:

$$U_v = U \cap N_U(v) \tag{2}$$

Table 1 The reference maximum clique algorithm REF

REF (U, C, S, S_{max})	REFCOL$(G(U), C)$
Input: A graph $G(V, E)$	Input: A subgraph $G(U)$
Ouput: A maximum clique in G in S_{max}	Output: A coloring $C(U)$; sorts U by color label
$U \leftarrow G, c(v_i) \leftarrow \min\{i, \Delta G\}, S, S_{max} \leftarrow \phi$	
1. Repeat Until $U = \phi$	1. sort vertices in U by non-increasing degree
2. $\mid$ $v \leftarrow \max_{v \in U}(c(v))$	2. SEQ(G, C)
3. $\mid$ $U \leftarrow U \setminus \{v\}$	3. sort vertices in U by non-increasing color in C
4. $\mid$ **If** $(\lvert S\rvert + c(v) \leq \lvert S_{max}\rvert)$ **Return**	
5. $\mid$ $S \leftarrow S \cup \{v\}$	
6. $\mid$ $U_v \leftarrow U \cap N_U(v)$	
7. $\mid$ **If** $(U_v = \phi)$	
$\mid$ $\mid$ **If**$(\lvert S\rvert > \lvert S_{max}\rvert)$ **Then** $\lvert S_{max}\rvert \leftarrow \lvert S\rvert$	
$\mid$ $\mid$ **Return**	
$\mid$ **endIf**	
8. $\mid$ REFCOL$(G(U_v), C_v)$	
9. $\mid$ REF (U_v, C_v, S, S_{max})	
10.$\mid$ $S \leftarrow S \setminus \{v\}$	
11. **endRepeat**	

Note that in every node all vertices in U_v are pairwise adjacent to every vertex in S_v and therefore valid candidates to enlarge it.

A tight upper bound for $\omega(G(U_v))$ is the size of vertex coloring ($\lvert C_v \rvert$) output of procedure REFCOL (table 1, right column). It is used to prune the search space in step 4 of REF (cf. [5] for a detailed explanation). REFCOL uses standard sequential heuristic coloring procedure SEQ: vertices are selected in a strict order as fixed on input and assigned the smallest possible label consistent with vertices already colored. A key advantage of using approximate coloring for bounding is that the color label $c(v_k)$ of any vertex $v_k \in U_v$ can be used as an upper bound on $\omega(G(W))$ where W is the subgraph with vertices lower than v_k (i.e. $W \subseteq U_v = \{v_1, v_2, \ldots, v_{k-1}\}$) as long as $c(v) \leq c(v_k) \; \forall v \in W$. In practice, REF always picks the candidate vertex with maximum color label to ensure this property (step 2), an important decision heuristic for the domain first described in [2].

3.1   Bit -Parallel MCP

An important recent algorithm for exact MCP is BBMC. BBMC employs bit strings to encode the domain so that main operators for child node computation (2) and SEQ coloring are reduced to bitmasks. BBMC also includes a number of subtle improvements such as *class coloring* implemented in the approximate coloring algorithm BBCOL and a fixed initial sorting of vertices by degree (cf. [5-6] for specific detailes).

Notation for bit string encodings used in this paper includes $BS(U)$ for vertex set U, $BS(v)$ for $N(v)$ (the neighbor set of v), and $BS[i]$, the induced subgraph by vertices encoded in the *i-th* bit block of BS. A subindex will be used when the encoded set needs to be made explicit (i.e. $BS_U[i]$).

In all cases the relative position of 1-bits inside the bit strings corresponds with the vertex index in the input graph as fixed initially. The adjacency matrix of the input graph is stored in memory as rows of bit strings $BS(v), \forall v \in V$.

4   Watched Subgraphs

This paper proposes to improve the efficiency of BBMC by *watching* two particular subgraphs of critical vertex sets (the ones containing the lowest and highest numbered vertices), and updating them on the fly during search. When this occurs, we also say the vertex set is *watched*.

In the case of MCP, watched critical sets allow for faster computation of the empty set condition for U_v in step 7 in REF as well as for the auxiliary sets used in approximate coloring (cf. [5-6]). Watched sets also allow for more efficient vertex reading from the compact bit strings. Moreover, critical computations for child node generation and coloring are simplified by reducing the range of meaningful bit mask operations.

The size of a watched subgraph is, at most, the register word size of the CPU (typically 64 in modern computers) so that any one of them can be encoded in a single bit block. Whenever a watched subgraph becomes empty, the nearest non empty subgraph becomes watched as a result of an appropriate update operation. The implementation requires two extra indexes (alias *sentinels*) s_l, s_h per watched set, which point to the watched subgraphs. Notation

$BS_U(s_l, s_h)$ is used to indicate that set U is being watched; $BS_U[s_l]$ and $BS_U[s_h]$ refer to the lower and upper subgraph respectively.

Table 2 Basic procedures INIT_WATCH, UPDATE, UPDATE_LOW and IS_EMPTY to operate with watched sets

INIT_WATCH($BS(s_l, s_h)$)
1. $s_l:$ $\begin{cases} \textbf{If}(BS \neq \phi)\ s_l \leftarrow \text{lowest non empty bitblock} \\ \textbf{If}(BS = \phi)\ s_l \leftarrow -1 \end{cases}$
2. $s_h:$ $\begin{cases} \textbf{If}(BS \neq \phi)\ s_h \leftarrow \text{hightest non empty bitblock} \\ \textbf{If}(BS = \phi)\ s_h \leftarrow -2 \end{cases}$

UPDATE($BS(s_l, s_h)$)	UPDATE_LOW($BS(s_l, s_h)$)
1. **If** ($s_l == -1 \parallel s_h == -2$) **Return**	1. **If** ($s_l == -1 \parallel s_h == -2$) **Return**
2. **While** ($BS[s_l] == \phi$)	2. **While** ($BS[s_l] == \phi$)
3. | $s_l \leftarrow s_l + 1$	3. | $s_l \leftarrow s_l + 1$
4. | **If** ($s_l > s_h$)	4. | **If** ($s_l > s_h$)
5. | | $s_l \leftarrow -1, s_h \leftarrow -2$	5. | | $s_l \leftarrow -1, s_h \leftarrow -2$
6. | | **Return**	6. | | **Return**
7. | **endIf**	7. | **endIf**
8. **endWhile**	8. **endWhile**
9. **While** ($BS[s_h] == \phi$)	
10.| $s_h \leftarrow s_h - 1$	
11. **endWhile**	

IS_EMPTY($BS(s_l, s_h)$)
1. **If** ($s_l > s_h$) **Return** TRUE **Else Return** FALSE

Table 2 shows the basic procedures implemented to operate with watched sets. Whenever a vertex set needs to be watched, INIT_WATCH is called to initialize the sentinels. Procedure UPDATE will typically be called after a critical operation which changes the set (i.e. child node computation step 6 in REF, or neighbor set removal during coloring (step 6 in BBCOLW, listed in table 3)) and updates the sentinels if required. Note that if both sentinels have the same index the watched set cannot be empty and all its vertices must belong to the pointed subgraph. Note also that UPDATE does not require updating the upper sentinel if empty set condition in step 4 of UPDATE is met. UPDATE_LOW is just a convenient case which only updates the lower sentinel. Finally, the empty set condition is evaluated by IS_EMPTY in constant time comparing both sentinels:

$$U = \phi \Leftrightarrow s_l > s_h, \ (s_l, s_h) \in BS(U) \tag{3}$$

Table 3 describes the use of watched sets in the new coloring procedure BBCOLW, which substitutes BBCOL in BBMC [cf. 5]. Both sets U and U' are watched so that terminating conditions for both the outer and inner loops are now computed by IS_EMPTY in constant time.

Table 3 The new coloring algorithm which uses watched sets

BBCOLW $(G(V,E), C, L)$
Input: A graph G to be colored
Output: $C(G)$, L (vertex set V ordered by color label)
$U \leftarrow V, U' \leftarrow V, L \leftarrow \phi, C \leftarrow \phi, i \leftarrow 1$
//initial step
INIT_WATCH$(BS(U'))$
INIT_WATCH$(BS(U))$
1. **Repeat Until** IS_EMPTY$(BS(U'))$
2. | **Repeat Until** IS_EMPTY$(BS(U))$
3. | | select the first vertex v from $BS_U[s_l]$
4. | | $C_i \leftarrow C_i \cup \{v\}$ //v is labeled with color i
5. | | $L \leftarrow L \cup \{v\}$ //output vertex list sorted by color
6. | | $U \setminus (\{v\} \cup N(v))$ in range $(BS_U[s_l], BS_U[s_h])$
7. | | UPDATE_LOW$(BS(U))$
8. | | $U' \leftarrow U \setminus \{v\}$ //removes colored vertex
9. | **endRepeat**
10.| UPDATE$(BS(U'))$
11.| $U \leftarrow U'$, $(s_l, s_h) \leftarrow (s_l', s_h')$ //copies also sentinels
12.| $i \leftarrow i+1$ //new color
13. **endRepeat**

5   Experiments

This section reports tests undertaken to evaluate the proposed watched set strategy (implemented as BBMCW) w.r.t. state of the art BBMC. The computer employed for the report is an Intel(R) Core(TM) i7 CPU 950 @ 3.07GHz with a 64-bit Windows O.S. and 6GB of RAM. It is the same as the one used in [6]; machine times for *dfmax* algorithm are available there (c.f. appendix section). In all experiments there was a time limit of 3600s and tests where averaged over 10 runs. We note that BBMC has been slightly modified to facilitate deployment and management w.r.t. versions described in [5-6] so times reported may differ slightly from those published elsewhere. We also note that only a small representative subset of all the experiments are published because of space constraints.

Table 4 reports results for a set of uniform random graphs (left) as well as a subset of the well known DIMACS graph benchmark [10] (right). Columns n, p refer to size and density of each graph; time is measured in seconds. The table shows that the impact of watched sets improves with size and density up to a 40% approx. In the random case, this is more acute for $n \geq 1000$. In the DIMACS graphs the differences in performance are smaller since all the graphs have less than 1000 nodes except two cases in the *phat* family. It is here that BBMCW performs best.

Table 4 User times for BBMC and BBMCW over a set of uniform random graphs and a subset of the DIMACS bechmark [10]. In bold best times for each test.

n	p	BBMC	BBMC	name	n	p	BBMC	BBMCW
300	0.5	**0.034**	0.036	*brock200_1*	200	0.745	0.249	**0.240**
300	0.6	**0.290**	0.297	*brock200_2*	200	0.496	0.002	0.002
300	0.7	**3.977**	4.148	*brock200_3*	200	0.605	0.009	0.009
500	0.4	0.117	**0.106**	*brock200_4*	200	0.658	0.043	**0.041**
500	0.5	0.937	**0.867**	*phat700-2*	700	0.498	2.777	**1.716**
500	0.6	15.599	**14.593**	*phat1000-2*	1000	0.49	150.610	**79.467**
100	0.3	0.511	**0.442**	*phat1500-1*	1500	0.253	2.605	**1.591**
100	0.4	6.696	**5.756**	*hamming10-2*	1024	0.99	0.019	**0.014**
100	0.5	156.732	**136.255**	*keller4*	171	0.649	0.008	**0.006**
300	0.1	0.371	**0.248**	*san200_0.7_2*	200	0.7	0.001	0.001
300	0.2	8.345	**5.819**	*san200_0.9_1*	200	0.9	**0.012**	0.015
500	0.1	2.894	**1.822**	*san200_0.9_2*	200	0.9	**0.034**	0.038
500	0.2	149.517	**104.387**	*sanr200_0.7*	200	0.702	0.097	**0.094**
100	0.1	57.356	**35.702**	*sanr400_0.5*	400	0.501	0.237	**0.216**
150	0.1	341.670	**226.890**	*sanr400_0.7*	400	0.7	71.685	**65.495**

6   Conclusions

This paper describes a novel idea of *watched subgraphs* to improve the performance of BBMC, a leading bit parallel solver for exact maximum clique. The idea is inspired in watched literals employed in efficient SAT solvers for Boolean constraint propagation. Reported results show that the efficiency of BBMC with watched sets improves as size and density of input graphs increase. Moreover, watched sets show comparable running times w.r.t. previous BBMC in the general case.

Acknowledgments. This work is funded by the Spanish Ministry of Science and Technology (ARABOT: DPI 2010-21247-C02-01) and supervised by CACSA whose kindness we gratefully acknowledge.

References

1. San Segundo, P., Rodríguez-Losada, D., Matía, F., Galán, R.: Fast exact feature based data correspondence search with an efficient bit-parallel MCP solver. Applied Intelligence 32(3), 311–329 (2010)
2. Tomita, E., Seki, T.: An efficient branch and bound algorithm for finding a maximum clique. In: Calude, C.S., Dinneen, M.J., Vajnovszki, V. (eds.) DMTCS 2003. LNCS, vol. 2731, pp. 278–289. Springer, Heidelberg (2003)
3. Konc, J., Janežič, D.: An improved branch and bound algorithm for the maximum clique problem. MATCH Commun. Math. Comput. Chem. 58, 569–590 (2007)
4. Tomita, E., Sutani, Y., Higashi, T., Takahashi, S., Wakatsuki, M.: A simple and faster branch-and-bound algorithm for finding a maximum clique. In: Rahman, M. S., Fujita, S. (eds.) WALCOM 2010. LNCS, vol. 5942, pp. 191–203. Springer, Heidelberg (2010)
5. San Segundo, P., Rodriguez-Losada, D., Jimenez, A.: An exact bit-parallel algorithm for the maximum clique problem. Computers & Operations Research 38(2), 571–581 (2011)
6. San Segundo, P., Matia, F., Rodriguez-Losada, D., Hernando, M.: An improved bit parallel exact maximum clique algorithm. Optimization Letters (2011), doi:10.1007/s11590-011-0431-y
7. Prosser, P.: Exact Algorithms for Maximum Clique: A Computational Study. Algorithms 5(4), 545–587 (2012)
8. BBMC release for 64 bit Win O.S. is available at:
   `http://www.intelligentcontrol.es/arabot/sites/default/`
   `files/frontpage/bbmc1.0.zip`
9. Moskewicz, M., Madigan, C., Zhao, Y., Zhang, L., Malik, S.: Chaff: engineering an efficient SAT solver. In: XXXVIII Proc. Annual Design Automation Conference (DAC 2001), pp. 530–535. ACM, New York (2001)
10. Johnson, D.S., Trick, M.A. (eds.): Cliques, coloring and Satisfiability. DIMACS Series in Discrete Mathematics and Theoretical Computer Science, vol. 26. American Mathematical Society, Providence (1996)

Decision Making and Optimization for Inspection Planning under Parametric Uncertainty of Underlying Models

Nicholas Nechval, Gundars Berzins, Vadim Danovich, and Konstantin Nechval

Abstract. Certain fatigued structures must be inspected in order to detect fatigue damages that would otherwise not be apparent. A technique for obtaining optimal inspection strategies is proposed for situations where it is difficult to quantify the costs associated with inspections and undetected failure. For fatigued structures, for which failures (fatigue damages) are only detected at the time of inspection, it is important to be able to determine the optimal times of inspection. Fewer inspections will lead to lower fatigue reliability of the structure upon demand, and frequent inspections will lead to higher cost. When there is a fatigue reliability requirement, the problem is usually to develop an inspection strategy that meets the reliability requirements. It is assumed that only the functional form of the underlying invariant distribution of time-to-failure is specified, but some or all of its parameters are unspecified. The invariant embedding technique proposed in this paper allows one to construct an optimal inspection strategy under parametric uncertainty. This strategy represents a sequence of inspection times satisfying the specific criterion, which takes into account the predetermined value of the conditional fatigue reliability of the structure. A numerical example is given.

Keywords: Fatigued structure, fatigue damage, parametric uncertainty, inspection strategy, optimization.

Nicholas Nechval · Gundars Berzins · Vadim Danovich
University of Latvia, EVF Research Institute, Statistics Department,
Raina Blvd 19, LV-1050 Riga, Latvia
e-mail: nechval@junik.lv

Konstantin Nechval
Transport and Telecommunication Institute, Applied Mathematics Department,
Lomonosov Street 1, LV-1019 Riga, Latvia
e-mail: konstan@tsi.lv

M. Ali et al. (Eds.): *Contemporary Challenges & Solutions in Applied AI*, SCI 489, pp. 123–129.
DOI: 10.1007/978-3-319-00651-2_17   © Springer International Publishing Switzerland 2013

1   Introduction

Many important fatigued structures (for instance, Transportation Systems and Vehicles: aircraft, space vehicles, trains, ships; Civil Structures: bridges, dams, tunnels; and so on) for which extremely high reliability is required are maintained by in-service inspections to prevent the reliability degradation due to fatigue damage. However, temporal transition of the reliability is significantly affected by the inspection strategy selected. Thus, to keep structures reliable against fatigue damage by inspections, it is clearly important in engineering to examine the optimal inspection strategy. In particular, it should be noticed that periodical inspections with predetermined constant intervals are not always effective, since a fatigue crack growth rate is gradually accelerated as fatigue damage grows, i.e. the intervals between inspections should be gradually smaller in order to restrain the reliability degradation by repeated inspections. Therefore, we need to construct the inspection strategy by paying attention to this case.

Barlow *et al.* [1] tackled this problem by assuming a known, fixed cost of making an inspection and a known fixed cost per unit time due to undetected failure. They then found a sequence of inspection times for which the expected cost is a minimum. Their results have been extended by various authors (Luss and Kander [2]; Sengupta [3]). Unfortunately, it is difficult to compute optimal checking procedures numerically, because the computations are repeated until the procedures are determined to the required degree by changing the first check time. To avoid this, Munford and Shahani [4] suggested a sub-optimal (or nearly optimal) but computationally easier inspection policy. This policy was used for Weibull and gamma failure distribution cases (Munford and Shahani [5]; Tadikamalla [6]). Numerical comparisons among certain inspection policies are given by Munford [7] for the case of Weibull failure times.

Most models, which are used for solving the problems of inspection planning, are developed under the assumptions that the parameter values of the models are known with certainty. When these models are applied to solve real-world problems, the parameters are estimated and then treated as if they were the true values. The risk associated with using estimates rather than the true parameters is called estimation risk and is often ignored.

In this paper, we consider the case when the functional form of the underlying invariant lifetime distribution is assumed to be known, but some or all of its parameters are unspecified. To make the discussion clear, we make the following restrictions on the inspection: (i) there is only one objective structure component of the inspection; (ii) if a fatigue crack is detected by the inspection, the component is immediately replaced with a new (virgin) one. To construct the optimal reliability-based inspection strategy in this case, the two criteria are proposed and the invariant embedding technique (Nechval *et al.* [8-9]) is used.

2   Planning Inspection Strategies under Certainty

In this paper we look at inspection strategies for items or structures that can be described as being in one of two states, one of which is preferable to the other. This preferred state might be described as 'working' whilst the other may represent some sort of 'failure'. The structures are originally known to be in a working state but may subsequently fail. In other words, at $t_0 = 0$ the structure is in state S_0 (working) but at a later time, t_1, the structure will move into state S_1 (failed). We suppose that we do not know when the transition from S_0 into S_1 will occur, and that a failure (fatigue crack) can only be detected through inspection. We deal with situations, where it is difficult to quantify the costs associated with inspections and undetected failure, or when these costs vary in time.

The inspection strategy defined is based on the conditional reliability of the structure. It is given as follows. Fix $0 < \gamma < 1$ and let

$$\tau_1 = \arg\left(\Pr\{X > \tau_1\} = \gamma\right), \tag{1}$$

$$\tau_j = \arg\left(\Pr\{X > \tau_j \mid X > \tau_{j-1}\} = \gamma\right), \quad j \geq 2, \tag{2}$$

where $\{\tau_j\}_{j=1, 2, \ldots}$ are inspection times, X is a random variable representing the lifetime of the structure. This is named as 'reliability-based inspection'. The above inspection strategy makes use of the information about the remaining life that is inherent in the sequence of previous inspection times. The value of γ can be seen as 'minimum fatigue reliability required' during the next period when the structure was still operational at last inspection time, that is, in other words, the conditional probability that the failure (fatigue crack) occurs in the time interval (τ_{j-1}, τ_j) without failure at time τ_{j-1} is always assumed $1 - \gamma$.

It is clear that if F_θ, the structure lifetime distribution with the parameter θ (in general, vector), is continuous and strictly increasing, the definition of the inspection strategy is equivalent to

$$\tau_j = \arg(\overline{F}_\theta(\tau_j) = \gamma^j), \quad j \geq 1, \tag{3}$$

or equivalent to

$$\tau_j^* = \arg\min_{\tau_j} [\overline{F}_\theta(\tau_j) - \gamma^j]^2, \quad j \geq 1, \tag{4}$$

where

$$\overline{F}_\theta(\tau_j) = 1 - F_\theta(\tau_j). \tag{5}$$

If it is known that each inspection costs c_1 and the cost of leaving an undetected failure (fatigue crack) is c_2 per unit time, then the total expected cost per inspection cycle is given by

$$E_\theta\{C\} = \sum_{j=1}^{\infty} \int_{\tau_{j-1}}^{\tau_j} [\, jc_1 + c_2(\tau_j - x)]f_\theta(x)dx$$

$$= c_1 \sum_{j=0}^{\infty} \overline{F}_\theta(\tau_j) + c_2 \sum_{j=1}^{\infty} \tau_j[\overline{F}_\theta(\tau_{j-1}) - \overline{F}_\theta(\tau_j)] - c_2 E_\theta\{X\}, \tag{6}$$

where $f_\theta(x)$ is the probability density function of the structure lifetime X,

$$E_\theta\{X\} = \int_0^{\infty} x f_\theta(x)dx. \tag{7}$$

Thus, we can choose γ such that $E_\theta\{C\}$ as defined in (6) is minimized.

3   Planning Inspection Strategies under Parametric Uncertainty

To construct the optimal reliability-based inspection strategy under parametric uncertainty, the two criteria are proposed.

The first criterion, which takes into account (3) and the past lifetime data of the structures of the same type, allows one to construct the inspection strategy given by

$$\tau_j = \arg(E_\theta\{\overline{F}_\theta(\tau_j)\} = \gamma^j), \quad j \geq 1, \tag{8}$$

where $\tau_j \equiv \tau_j(\hat{\theta})$, $\hat{\theta}$ represents either the maximum likelihood estimator of θ or the sufficient statistic S for θ, i.e., $\tau_j \equiv \tau_j(S)$. This criterion is named as 'unbiasedness criterion'.

The second criterion (preferred), which takes into account (4) and the past lifetime data of the structures of the same type, allows one to construct the inspection strategy given by

$$\tau_j^* = \arg\min_{\tau_j} E_\theta\{[\overline{F}_\theta(\tau_j) - \gamma^j]^2\}, \quad j \geq 1, \tag{9}$$

This criterion is named as 'minimum variance criterion'.

It will be noted that in practice, under parametric uncertainty, the criterion,

$$\tau_j = \arg(\overline{F}_{\hat{\theta}}(\tau_j) = \gamma^j), \quad j \geq 1, \tag{10}$$

is usually used. This criterion is named as 'maximum likelihood criterion',

To find a sequence of inspection times, $\tau_j \equiv \tau_j(\hat{\theta})$ or $\tau_j \equiv \tau_j(S)$, $j \geq 1$, satisfying either (8) or (9), the invariant embedding technique (Nechval et al. [8-9]) can be used. Let us assume that each inspection costs c_1 and the cost of leaving an undetected failure (fatigue crack) is c_2 per unit time, then under parametric uncertainty we can choose γ such that $E_\theta\{E_\theta\{C\}\}$ is minimized.

3.1 Inspection Strategies for the Exponential Lifetime Distribution

Let $X_1, \ldots, X_n$ be the random sample of the past independent lifetime observations from the fatigued structures of the same type, which follow the exponential distribution with the probability density function

$$f_\theta(x) = (1/\theta)\exp(-x/\theta), \quad x \geq 0, \quad \theta > 0, \tag{11}$$

where the parameter θ is unknown. It can be shown that the reliability-based inspection strategies for a new fatigued structure of the same type are given as follows.

The unbiased inspection strategy (UIS):

$$\tau_j = [\gamma^{-j/n} - 1]S, \quad j \geq 1. \tag{12}$$

The minimum variance inspection strategy (MVIS):

$$\tau_j^* = [1 - \gamma^{j/(n+1)}][2\gamma^{j/(n+1)} - 1]^{-1} S, \quad j \geq 1, \tag{13}$$

where $S = \sum_{i=1}^{n} X_i$ is the sufficient statistic for θ.

The maximum likelihood inspection strategy (MLIS):

$$\tau_j = j(S/n)\ln\gamma^{-1}, \quad j \geq 1. \tag{14}$$

3.2 Optimization of γ for the Unbiased Inspection Strategy

Theorem 1. Let us assume that each inspection of the unbiased inspection strategy costs c_1 and the cost of leaving an undetected failure (fatigue crack) is c_2 per unit time, then γ minimizing $E\{E_\theta\{C\}\}$ is given by

$$\gamma^* = \arg\left(\left[\frac{1 - \gamma^{(n+1)/n}}{1 - \gamma}\right]^2 \left[\frac{1 - \gamma^{-1/n} + (\gamma^{-1} - 1)/n}{\gamma^{1/n}}\right] = \frac{n+1}{n}\frac{c_1}{c_2}\frac{1}{S}\right). \tag{15}$$

Proof. Taking into account (6) and (12), we have

$$\gamma^* = \operatorname*{argmin}_\gamma E_\theta\{E_\theta\{C\}\} = \operatorname*{argmin}_\gamma E_\theta\left\{c_1\sum_{j=0}^{\infty}\overline{F}_\theta(\tau_j) + c_2\sum_{j=1}^{\infty}\tau_j[\overline{F}_\theta(\tau_{j-1}) - \overline{F}_\theta(\tau_j)] - c_2 E_\theta\{X\}\right\}$$

$$= \arg\left(\left[\frac{1 - \gamma^{(n+1)/n}}{1 - \gamma}\right]^2 \left[\frac{1 - \gamma^{-1/n} + (\gamma^{-1} - 1)/n}{\gamma^{1/n}}\right] = \frac{n+1}{n}\frac{c_1}{c_2}\frac{1}{S}\right). \tag{16}$$

This ends the proof. ∎

4   Numerical Example

Let $X_1, \ldots, X_n$ be the random sample of the past independent lifetime observations from the same fatigued structures, which follow the exponential distribution (11), where $n = 2$ and the parameter θ is unknown. The sufficient statistic for θ is $S = 335$ hours. In order to construct the reliability-based inspection strategy for a new fatigued structure of the same type, the unbiasedness criterion (8) will be used. Let us assume that each inspection of the UIS costs $c_1=1$ (in terms of money) and the cost of leaving an undetected failure (fatigue crack) is $c_2=2$ ((in terms of money)) per unit time. Then it follows from (15) that $\gamma^* =0.95$. Fig.1 depicts the relationship between $E_\theta\{E_\theta\{C\}\}/\theta$ and γ.

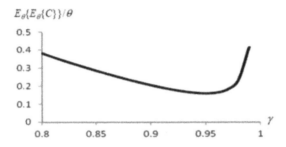

Fig. 1 Relationship between $E_\theta\{E_\theta\{C\}\}/\theta$ and γ

The optimal inspection times (in terms of hours) for the unbiased inspection strate-gy are given in Table 1.

Table 1 Optimal inspection times (in terms of hours) for the unbiased inspection strategy

j	1	2	3	4	5	6	7	8	9	...
τ_j	8.70	17.63	26.79	36.19	45.83	55.73	65.88	76.29	86.98	...

We have illustrated the technique of constructing inspection strategies for the exponential lifetime distribution. Application to other log-location-scale distributions could follow directly.

References

1. Barlow, R.E., Hunter, L.C., Proschan, F.: Optimum Checking Procedures. Journal of the Society for Industrial and Applied Mathematics 11, 1078–1095 (1963)
2. Luss, H., Kander, Z.: Inspection Policies when Duration of Checkings is Non-negligible. Operational Research Quarterly 25, 299–309 (1974)

3. Sengupta, B.: Inspection Procedures when Failure Symptoms are Delayed. Operational Research Quarterly 28, 768–776 (1977)
4. Munford, A.G., Shahani, A.K.: A Nearly Optimal Inspection Policy. Operational Research Quarterly 23, 373–379 (1972)
5. Munford, A.G., Shahani, A.K.: An Inspection Policy for the Weibull Case. Operational Research Quarterly 24, 453–458 (1973)
6. Tadikamalla, P.R.: An Inspection Policy for the Gamma Failure Distribution. Operational Research Quarterly 30, 77–78 (1979)
7. Munford, A.G.: Comparison among Certain Inspection Policies. Management Science 27, 260–267 (1981)
8. Nechval, N.A., Berzins, G., Purgailis, M., Nechval, K.N.: Improved Estimation of State of Stochastic Systems via Invariant Embedding Technique. WSEAS Transactions on Mathematics 7, 141–159 (2008)
9. Nechval, N.A., Nechval, K.N., Purgailis, M.: Prediction of Future Values of Random Quantities Based on Previously Observed Data. Engineering Letters 9, 346–359 (2011)

Topological Feature Mining for Rambling Activities

Masakatsu Ohta and Miyuki Imada

Abstract. A method for investigating rambling activities of moving objects is proposed. The goal is to construct common metrics used in various environments for characterizing the trajectory followed by rambling objects. Rambling activities are multi-stop, multi-purpose trips with trajectories with many intersections. Mathematical knot theory is introduced to examine the topological relation between intersections. The trajectories in an environment are represented in a vector space consisting of prime knots. Like a prime number, a prime knot is universal; thus, it is possible to compare the features of rambling activities across environments. An experiment using real-world taxi trajectories demonstrated that our method effectively classifies rambling activities according to daytime, nighttime, and a special event.

1 Introduction

In recent years, many efforts to design a space in which people are induced to ramble (multi-stop and multi-purpose trips) have been made. To avoid urban sprawl and to revitalize downtowns, urban planning are aimed at enabling comfortable movement around downtowns [9]. Many shopping malls have various non-shopping facilities such as restaurants and movie theaters. People spend much time in these facilities. Consequently, they visit many stores and buy a large amount of merchandise [11].

Rambling activities have been represented by computational models. A pedestrian's next positions are predicted by a mixed Markov chain [2]. Using a large amount of locations collected using positioning methods, such as global positioning systems (GPSs), data mining approaches extract the routes frequently followed by moving objects and predict the next location of a moving object by sequence analysis such as the apriori algorithm [5, 7].

Masakatsu Ohta · Miyuki Imada
NTT Network Innovation Laboratories, NTT Corporation,
3-9-11 Midori-cho, Musashino-shi, Tokyo, 180-8585, Japan
e-mail: {ohta.masakatsu,imada.miyuki}@lab.ntt.co.jp

M. Ali et al. (Eds.): *Contemporary Challenges & Solutions in Applied AI*, SCI 489, pp. 131–139.
DOI: 10.1007/978-3-319-00651-2_18 © Springer International Publishing Switzerland 2013

As is done in business process benchmarking [3], a cross case comparison of the features of rambling activities in other environments and fields is effective for designing ideal spaces. By comparing the best practices that have attained ideal performances among environments that are similar in properties, such as geographical conditions, planners may conceive innovative solutions to achieve their aims. However, previous studies are not suitable for this comparison. They were focused on the predicting behavior of moving objects only in each environment and have not presented any framework for quantitative comparison of these behaviors across environments. Hence, it is impossible to answer questions such as "In what cities do people move around as much as they do in my city?" The problem is one of constructing metrics used in various environments for characterizing the trajectory followed by rambling objects.

We propose a method for investigating the topological feature of trajectories caused by the rambling activities of moving objects. Topological relations between the intersections that an object repeatedly visits in an environment are represented by primitive factors available in different environments.

2 Topological Feature of Rambling Activities

2.1 Relation between Intersections

Revisiting the same places is a notable feature of rambling activities. Thus, we address the rambling activities of moving objects that repeatedly visit the same places in an environment. In contrast, we are not concerned here with those that cause trajectories without any intersections.

The more an object moves around an area, the more likely the trajectory followed by the object will intersect itself. To characterize a trajectory with many intersections, we examine how they connect with each other. Mathematical knot theory is introduced to compare this topological feature across different environments.

2.2 Knot Theory

Mathematical knot theory is used to investigate the topological structure of a closed curve in three dimensional Euclidean space that does not intersect itself anywhere [1]. In the proposed method, a trajectory corresponds to a knot. Transforming a trajectory into a knot is described in the next section. The original knot and the deformations of that curve through space without permitting the curve to pass through itself are equivalent. The simplest knot is an unknotted circle and called the trivial knot or the unknot. For each knot, there is a mirror image obtained from the original knot by reflecting it on some plane (Fig.1). The mirror image is not necessarily equivalent to the original knot. Each diagram in Fig. 1 represents a projection onto a 2-dimensional plane of the knot and is called a knot diagram. The points at which a knot intersects itself in a knot diagram are called crossings. The crossing number of a knot denotes the minimal number of crossings of that equivalent knot.

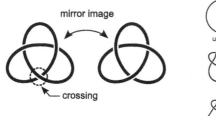

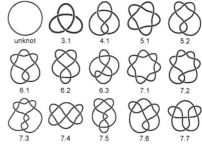

Fig. 1 Knot and its mirror image. Knots were drawn using KnotPlot [10].

Fig. 2 Knot diagram of prime knots up to seven crossings. Mirror images are not shown.

It is possible to draw a diagram in which two knots are connected without creating new crossings. The resulting knot is called a composition of the two knots and this operation is denoted by #. If a knot is not the composition of any two nontrivial knots, it is called a prime knot. Any composite knot is divided into a unique set of prime knots. Prime knots have been catalogued in terms of the crossing number (Fig.2)[6, 8].

To identify prime knots, we use the HOMFLY polynomial [4], which is a knot invariant. It takes the same value for equivalent knots, but not vice versa. The HOMFLY polynomial for the left knot in Fig. 1 is $P = -v^4 + 2v^2 + z^2v^2$, and that for the right one is $P = -v^{-4} + 2v^{-2} + z^2v^{-2}$, where v and z are variables. The polynomial of the unknot $P(unknot)$ is defined as 1. Regarding the HOMFLY polynomial of a composition of two knots K_1 and K_2,

$$P(K_1\#K_2) = P(K_1)P(K_2) \tag{1}$$

holds. A method of computing the HOMFLY Polynomial is beyond the scope of this paper. For further details, see [1, 4].

2.3 Feature Representation

By dividing the HOMFLY polynomial of a knot by those of prime knots until the quotient is 1 by using Eq. 1, we obtain prime knots as elements of the original knot and how many prime knots compose it. Thus, by decomposing all knots formed from trajectories in an environment into prime knots, we obtain a histogram of prime knots as elements of the knots corresponding to the trajectories. This histogram represents the topological feature of the rambling activities in an environment, and its entries are used as the feature vector. This denotes that a set of prime knots is used as a basis of the vector space. Like a prime number, a prime knot is universal; thus, to compare the topological features across different environments, a common set of prime knots is used in different environments. Since the number of trajectories depends on environments, the norm of a feature vector is normalized to one.

Moreover we use an alternative feature vector corresponding to the histogram of the crossing numbers calculated from that of prime knots. The dimension of this vector is suppressed to the maximum crossing numbers of the prime knot set.

3 Forming Knot from Trajectory

In this section we describe the transformation of a 2-dimensional trajectory (x,y) into a 3-dimensional knot (x,y,z). If this knot is not the unknot, it is assumed that the moving object has rambled around in the environment.

3.1 Grade Separation

Consider a three-dimensional curve with the same x and y values of a trajectory. This curve is called a trajectory curve. Every time a trajectory intersects itself, the z value of the trajectory curve's point related to that intersection is coordinated to prevent the curve from intersecting itself. A trajectory is permitted to intersect at the same places many times. The z value relating to the i-th intersection is calculated as follows:

$$z_i = \begin{cases} \max A_i + 1 & \text{if } \sigma > 0 \\ \min A_i - 1 & \text{otherwise,} \end{cases} \tag{2}$$

where $\sigma = (-1)^i$, and $A_i = \{z \mid x = x_i, y = y_i, (x,y,z) \in C_i\}$. C_i is the trajectory curve before the i-th intersection. For the points on a trajectory curve unrelated to intersections, their z values are set to zero. Along a trajectory, a trajectory curve grows by alternating over and under every intersection, like knitting.

3.2 Closing Curve

Assuming that a trajectory is not closed, the new start and end points are created outside the rectangular solid including the trajectory curve mentioned in Section 3.1, and connected outside that rectangular solid. The arbitrary closings outside the rectangular solid create the equivalent knots.

The new start point is created by drawing a line in the positive direction of the Z axis from the original start point. The new end point is created in the same manner except that the direction of drawing the line equals the sign of σ at the last intersection (Fig.3).

3.3 Dynamic Relaxation

Some routes in a trajectory often overlap. As for the parts of the trajectory curve relating to the overlapping routes, the z values are first set to zero. To remove the overlap in a trajectory curve, a dynamic relaxation is applied to the curve. This relaxation changes a knot into its equivalents with fewer crossings.

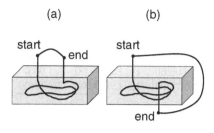

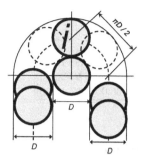

Fig. 3 Closing curve for σ at last intersection; (a) $\sigma > 0$, (b) $\sigma < 0$

Fig. 4 Collision avoidance. Particle i behaves as hard sphere for those denoted as solid circles.

Let us assume that a trajectory curve is a chain of particles successively connected by springs. The elastic energy of this chain is represented as

$$E \propto \sum_i (|\boldsymbol{q}_{i-1} - \boldsymbol{q}_i|^2 + |\boldsymbol{q}_{i+1} - \boldsymbol{q}_i|^2), \qquad (3)$$

where $\boldsymbol{q}_i$ is the coordinate of the i-th particle. By updating $\boldsymbol{q}_i$ to decrease Eq. 3, particles get close to each other and a trajectory curve shrinks over time. Since an overlapping particle is pulled by those whose z values are nonzero, the overlap is removed.

During deformation, the curve is forbidden to pass through itself. To satisfy this constraint, we avoid the collision of particles (Fig. 4). The i-th particle behaves as a hard sphere for the particle set: $K_i = \{k \mid l_{ik} > \pi D/2, d_{ik} < D, k \neq i\}$, where D is the diameter of particles, l_{ik} is the minimum length between the i-th and k-th particles through the curve, and d_{ik} is the linear distance between the two particles. To satisfy the constraint, $\boldsymbol{q}_i$ is updated using the following equation:

$$\Delta \boldsymbol{q}_i = \frac{1}{|K_i|} \sum_{k \in K_i} \left\{ (D - |\boldsymbol{q}_i - \boldsymbol{q}_k|) \frac{(\boldsymbol{q}_i - \boldsymbol{q}_k)}{|\boldsymbol{q}_i - \boldsymbol{q}_k|} \right\}. \qquad (4)$$

For this constraint, a trajectory curve stops shrinking at a certain size.

Before the dynamic relaxation, the z values of a trajectory curve are multiplied by D. The dynamic relaxation is executed in the following steps. First, some particles are interpolated into a trajectory curve at intervals of $D/2$ so that the constraint effectively operates. Second, $\boldsymbol{q}_i$ is updated according to Eq. 4, and except for the updated particles by Eq. 4, $\boldsymbol{q}_i$ is updated to decrease Eq. 3 using the steepest descent method. These steps are repeated until the length of the trajectory curve converges.

Table 1 Time period

index	period	index	period	index	period
1	2/2 18:00 – 2/3 06:00	5	2/4 18:00 – 2/5 06:00	9	2/6 18:00 – 2/7 06:00
2	2/3 06:00 – 2/3 18:00	6	2/5 06:00 – 2/5 18:00	10	2/7 06:00 – 2/7 18:00
3	2/3 18:00 – 2/4 06:00	7	2/5 18:00 – 2/6 06:00	11	2/7 18:00 – 2/8 06:00
4	2/4 06:00 – 2/4 18:00	8	2/6 06:00 – 2/6 18:00	12	2/8 06:00 – 2/8 18:00

4 Experiment

4.1 Dataset

We evaluated the performance of our method on a T-Drive trajectory dataset that
contains the GPS trajectories of 10357 taxis from February 2 to February 8, 2008
in Beijing [12, 13]. The Chinese New Year (Spring Festival) of 2008, which was on
February 7, is included. Each record consists of the taxi index, time, and location.
The customer index is not included.

4.2 Analysis

It seems reasonable to assume that people change their behavior according to
daytime, nighttime, and special events, and these changes are reflected in the taxi
trajectories. The trajectories of 1500 taxis were selected and divided into 12 time
periods (Tab. 1). Each period was regarded as one environment. The topological
feature of the trajectories in each period were obtained and the difference between
daytime (6:00 - 18:00) and nighttime (18:00 - 6:00) and the effect of the Spring Festival on the rambling activities were examined. A taxi driver selects the best route for
each customer, and there are hardly any intersections in that trajectory. However, the
trajectory of a taxi within a certain period includes some customer trips; therefore,
it is a multi-stop, multi-purpose trip chain and considered to have characteristics of
rambling activities.

The analysis range was set to L(km). Next, a part of a trajectory within a square
of the analysis range was extracted. The center of the square was adjusted to the
centroid of each trajectory. By taking into account measurement error, the extracted
trajectory was approximated with the square mesh of side $L/20$ and transformed into
a knot. The successive points on a trajectory within the mesh are considered as the
same points. Thus, the diameter of the particles in the dynamic relaxation was set
to this mesh size. To examine the effect of periods on the rambling activities, L was
set to the same value in all periods and obtained using median X-range and Y-range
of the trajectories in all periods, which removes trajectories with extremely long
lengths. As a result, L was 20 km.

A set of the prime knots up to 12 crossings was used as the basis of the vector
space. The number of prime knots with the unique HOMFLY polynomials was 5214.

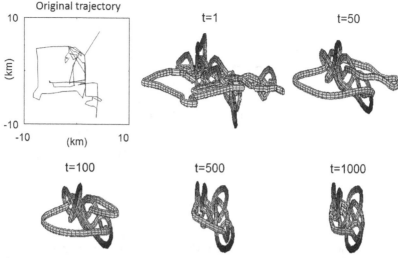

Fig. 5 Dynamic relaxation process

4.3 Results

The process of dynamic relaxation is shown in Fig. 5. A knot representing a trajectory shrinks and changes into an equivalent one with fewer crossings over time.

The feature vector created by the crossing numbers of the prime knots found in each period is shown in Fig. 6. The crossing numbers are from 3 to 12; thus, the dimension of the vector is 10. The effect of the Spring Festival is especially noticeable

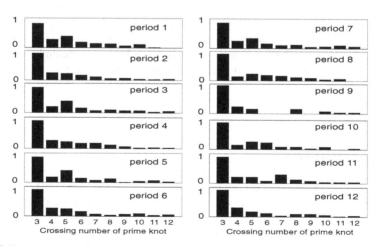

Fig. 6 Feature vector of period

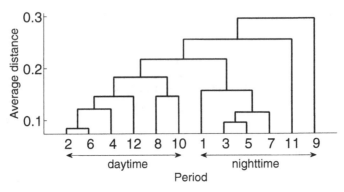

Fig. 7 Hierarchical cluster tree

in Period 9. Knots with small crossing numbers were frequently observed. From the point of view of rambling activities, the taxi trajectories became simple.

Using these feature vectors, hierarchical clustering was used to examine the difference between periods (Fig. 7). Except for the periods around the Spring Festival, daytime and nighttime are clearly distinguished. Periods 9 and 11 show particular patterns that belong to neither daytime nor nighttime; thus, the Spring Festival had a stronger effect during nighttime.

5 Conclusion

We proposed a method for investigating the topological feature of trajectories caused by rambling objects. Based on mathematical knot theory, trajectories in an environment are represented in vector space consisting of prime knots. Like a prime number, a prime knot is universal; thus, the topological feature of trajectories in an environment is compared with those of trajectories in different environments. An experiment using real-world taxi trajectories indicated that our method effectively classifies persons' rambling activities according to the environment.

References

1. Adams, C.C.: The Knot Book: An Elementary Introduction to the Mathematical Theory of Knots. American Mathematical Society (2004)
2. Asahara, A., Maruyama, K., Sato, A., Seto, K.: Pedestrian-movement prediction based on mixed Markov-chain model. In: Proceedings of the 19th ACM SIGSPATIAL International Conference on Advances in Geographic Information Systems, pp. 25–33. ACM Press (2011)
3. Boxwell, R.J.: Benchmarking for Competitive Advantage. McGraw-Hill Professional Publishing (1994)

4. Freyd, P., Yetter, D., Hoste, J., Lickorish, W., Millett, K., Ocneanu, A.: A new polynomial invariant of knots and links. Bulletin of the American Mathematical Society 12(2), 239–247 (1985)

5. Monreale, A., Pinelli, F., Trasarti, R., Giannotti, F.: WhereNext: a Location Predictor on Trajectory Pattern Mining. In: Proceedings of the 15th ACM SIGKDD International Conference on Knowledge Discovery and Data Mining, pp. 637–646. ACM Press (2009)

6. Morrison, S., Bar-Natan, D.: Knot Atlas, `http://katlas.org/wiki/Main_Page` (cited February 8, 2013)

7. Morzy, M.: Mining Frequent Trajectories of Moving Objects for Location Prediction. In: Perner, P. (ed.) MLDM 2007. LNCS (LNAI), vol. 4571, pp. 667–680. Springer, Heidelberg (2007)

8. Rolfsen, D.: Knots and Links. AMS Chelsea Publishing (1976)

9. Saunders, W.S.: Urban Planning Today: A Harvard Design Magazine Reader. Univ Of Minnesota Press (2006)

10. Scharein, R.: KnotPlot, `http://www.knotplot.com` (cited February 8, 2013)

11. Underhill, P.: Call of the Mall: The Geography of Shopping by the Author of Why We Buy. Simon & Schuster (2004)

12. Yuan, J., Zheng, Y., Xie, X., Sun, G.: Driving with knowledge from the physical world. In: Proceedings of the 17th ACM SIGKDD International Conference on Knowledge Discovery and Data Mining, vol. 5, pp. 316–324. ACM Press (2011)

13. Yuan, J., Zheng, Y., Zhang, C., Xie, W., Xie, X., Sun, G., Huang, Y.: T-drive: driving directions based on taxi trajectories. In: Proceedings of the 18th SIGSPATIAL International Conference on Advances in Geographic Information Systems, GIS 2010, pp. 99–108. ACM (2010)

Part VII
Pattern Recognition

Confusion Matrix Based Reweighting

Vincent Damian Warmerdam and Zoltán Szlávik

Abstract. This paper introduces a method to rebalance the output of classification algorithms using the corresponding confusion matrices. This is done by modifying the classification output, i.e. reweighting predictions, when they can be interpreted as probabilities. The method is evaluated and analyzed via experiments involving a number of classifiers and both standard and real life datasets. Our results show that confusion matrix based reweighting can be used to achieve certain kinds of balance in classification, while maintaining the same level of accuracy.

1 Introduction

When applying classification algorithms, one often needs to deal with the so called class imbalance problem. For instance, if 99% of the data describes a certain class, e.g. "not fraud", that is a completely different situation from having a balanced dataset in which all classes are represented with close to equal frequency. In addition to this class imbalance problem at the level of the dataset (i.e., the *input*), however, one might also face a similar class imbalance at the *output* level.

Several approaches exist to tackle the class imbalance problem. Some of these focus on creating a balance at input level, while some attempt to change classification output. Naturally, these methods are not concerned directly with class balance at output level, but it is implicitly assumed that a balanced input will result in an balanced output. [1], [2], [3]

A widely used approach to manipulate classification balance is the utilization of cost matrices. Using a cost matrix, one can achieve certain balance in the distribution of the predicted classes, or even manipulate the predictions in order to optimize on certain evaluation measures. Cost matrices are often

Vincent Damian Warmerdam · Zoltán Szlávik
Department of Computer Science, VU University Amsterdam

M. Ali et al. (Eds.): *Contemporary Challenges & Solutions in Applied AI*, SCI 489, pp. 143–148.
DOI: 10.1007/978-3-319-00651-2_19   © Springer International Publishing Switzerland 2013

determined by domain experts, but determining the exact values of costs is not always trivial. Cost matrices can also be designed to produce the same output class distribution as that of the input. [4], [5], [6].

In this paper, we focus on the class imbalance problem at the output level. Our proposed approach, described in the next section, is in essence a new way to create cost matrices. However, it comes without the need for specific expert knowledge, but with a direct use of the quality of the predictions.

2 Confusion Matrix Based Reweighting

We will now explain the proposed method more formally and in-depth. We are given a supervised machine learning algorithm that attempts to classify vector y using dataset X. The algorithm outputs probabilities such that the class with the highest probability will be selected as the predicted class. The training dataset contains n points and k classes such that the predicted probabilities are summarized in a $n \times k$ matrix P^{old}. If $P_i(j')$ is the probability that the algorithm assigns point i to class j then P^{old} is given by;

$$P^{old} = \begin{bmatrix} P_1(1') & P_1(2') & .... & P_1(5') & ... & P_1(k') \\ \vdots & \vdots & \vdots & \vdots & ... & \vdots \\ P_n(1') & P_n(2') & .... & P_n(5') & ... & P_n(k') \end{bmatrix}$$

These probabilities then give us the predictions $\hat{y}$ by selecting the class with the highest probability.

$$\hat{y} = \begin{bmatrix} argmax\{P_1(1'), ..., P_1(k')\} \\ \vdots \\ argmax\{P_n(1'), ..., P_n(k')\} \end{bmatrix}$$

From this output we can construct a confusion matrix CFM that counts the number of predictions against its true values.

$$CFM = \begin{bmatrix} n_{1|1'} & n_{1|2'} & .... & n_{1|j'} & ... & n_{1|k'} \\ \vdots & \vdots & \vdots & \vdots & ... & \vdots \\ n_{k|1'} & n_{k|2'} & .... & n_{k|j'} & ... & n_{k|k'} \end{bmatrix}$$

where $n_{a|b}$ counts the number of times the algorithm assigns class b when it's true class is class a. From this matrix we can calculate $P(a|b')$, the probability that a point belongs to class a while the algorithm predicted it to be from class b. This probability is calculated by;

$$P_{a|b'} = \frac{n_{a|b'}}{\sum_{j=1}^{k} n_{j|b'}}$$

By doing this for every true class and for every predicted class we get a $k \times k$ matrix M_S;

$$M_S = \begin{bmatrix} P_{1|1'} & P_{1|2'} & .... & P_{1|j'} & ... & P_{1|k'} \\ \vdots & \vdots & \vdots & \vdots & ... & \vdots \\ P_{k|1'} & P_{k|2'} & .... & P_{k|j'} & ... & P_{k|k'} \end{bmatrix}$$

Matrix M_S can be interpreted as a bayesian filter: using conditional probabilities we attempt to improve accuracy in our model. We can apply the law of total probability at a large scale. Here as P^{old} is a $n \times k$ matrix and M_S is a $k \times k$ matrix. According to this law the true probability can be determined by the following matrix operation.

$$P_{new} = P_{old} M_S$$

To prevent the matrix M_S from creating very drastic changes one could introduce a scaling parameter α_{BAYES} that scales the new output to be a combined value from the old output and the new output after the bayesian filter has been applied.

$$P_{new} = \alpha_{BAYES} P_{old} M_S + (1 - \alpha_{BAYES}) P_{old}$$

Notice that if the original confusion matrix is a diagonal matrix that this method will not cause any change in the classification output. If the matrix contains more non diagonal elements it will cause a change in classification.

3 Experimental Setup

In order to investigate the confusion matrix based reweighting method, we performed several experiments with various datasets, classifiers and their versions, as well as with various levels of the smoothing parameter α. We used the statistical package R[1] this.

We used six datasets. Four of them were taken from the UCI machine learning repository[2], while two were real life datasets (Studyflow) about course profiles of Dutch secondary school students.

The four classifiers used were chosen because of their probabilistic output and wide usage. We used **logistic regression (LR)** based on the *glm* R package, **feed forward neural network (NN)** based on the *nnet* package, **support vector machine (SVM)** based on the *e1071* package and **random forest (RANDF)** based on the *randomForest* package.

We experimented with all classifier-dataset combinations. For logistic regression, neural networks and random forests we also experimented with the iterations allowed to find the optimal weights in order to see if this affects

---

[1] http://www.r-project.org/
[2] http://archive.ics.uci.edu/ml/

the results. We also investigated various values of α to measure what is the effect on the precision/recall ratio as well as on model accuracy.

For each run, we acquired the confusion matrix based on the training set, from this the reweighing matrix was derived which rescales the output based on α.

4   Results

In this section, we present two main sets of results. First, we illustrate how our method works using the Abalone dataset, then we provide a comparison between the performance of various classifier-dataset-α combinations.

4.1   *Detailed Results on a Dataset*

To illustrate the behavior of our method, in this subsection we present our results on the Abalone dataset. In this dataset, the algorithm needs to classify a californian sea creature to be male (M), female (F) or an infant (I) based on body measurements.

Applying our method with $\alpha = 1$ seem to "flip" the confusion matrix along the diagonal (see Table 1) which shows the significant effect of reweighting. In general, the use of the M_S matrix allows to recognize mistakes being made in the original assignment of prediction probabilities. This, however, causes the algorithm to make less of one type of mistakes but it will cause it to make new ones. As the accuracy remained comparable, and because of the "flipping" effect observed.

The workings of our method are further illustrated in Figure 1 where we display classification performance as a function of α (this time using 30 runs of a neural network). As α increases, the algorithm favors one class (F) more than another (M), while the third class (I) does not seem to be affected.

To sum up, we have seen that it is possible to set the value of parameter α so that performance associated with the two difficult classes is in balance, while the total accuracy of classification does not change considerably.

Confusion matrix based reweighing caused the same behavior across the other datasets. If the machine learning algorithm showed good performance (i.e., the confusion matrix contained only diagonal elements) the classification would not change by applying our method.

4.2   *Results over Various Datasets and Classifiers*

Regarding balance, for all datasets we have been able to find a value for α (other than zero) such that the classification became balanced. The average values of the optimal α values in terms of class balance (i.e. when total false positives = total false negatives) is shown in Table 2.

Table 1 Confusion matrices for a support vector machine applied to the abalone data. Both the normal confusion matrix of the original SVM ($\alpha = 0$, left) and the SVM with our filter ($\alpha = 1$, right) are shown.

	SVM F	SVM I	SVM M	SVM-F F	SVM-F I	SVM-F M
true F	196	371	92	242	11	179
true I	9	112	549	324	104	402
true M	137	454	168	93	555	178

Table 2 The table that lists at what the average value of α where the number of false positives equals the number of false negatives. The TESTS and COURSES datasets are based on 100 algorithm runs, all other datasets results are based on 1000 algorithm runs. We did not include the CANCER datasets as there was no effect of the method there.

	HANDW	ABALONE	TICTAC	CARDIO	TESTS	COURSES
LR-	0.4728	0.5325	*NE*	0.0342	0.3096	0.4132
LR+	0.3740	0.5464	*NE*	0.0354	0.3808	0.3600
NNET-	0.5302	0.7301	*NE*	0.1015	0.4720	0.5302
NNET+	0.3310	0.6565	*NE*	0.1020	0.3648	0.3310
RANDF-	0.0983	0.5846	0.2360	0.0283	0.4864	0.6072
RANDF+	0.3158	0.8016	0.4720	0.0284	0.6192	0.9216
SVM	*NE*	0.7200	*NE*	*NE*	*NE*	0.4282

What we see in this table is that between algorithms and datasets there does not seem to be a clear pattern. A dataset is classified differently by each algorithm and as of such the imbalance in the output is different. For all method-dataset combinations our method was able to balance the output if it was imbalanced with our method.

5 Discussion

In this paper,we have investigated the use of confusion matrices, and found that using this method it is possible to rebalance classification output, without damaging the output too much. Confusion matrix based reweighting offers an alternative to various other methods, e.g. sampling techniques and cost matrices, that are to tackle aspects of imbalance. Using the α parameter we can directly influence the balance, and potentially set it to a value that results in a desired output balance. Our method is computationally cheap, as it only requires a confusion matrix, and according to our results the confusion matrix on the training set already is promising to be used for balancing.

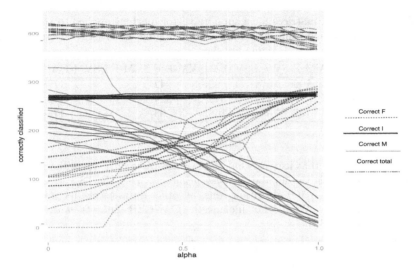

Fig. 1 The accuracy for 30 neural networks applied on the Abalone dataset for increasing values of α. In this setting we can notice how the algorithm has no problem identifying the infant class but it finds it hard to distinguish between male and female. While the total accuracy remains relatively stable, prior accuracy of the algorithm (i.e. $\alpha = 0$, equivalent to our method not applied) seems to favor either the M or F classes in the beginning over each other.

References

1. Garcia, et al.: The class imbalance problem in pattern classification and learning. Congreso Espanol de Informatica, 283–291 (2007)
2. Galar, et al.: A Review on Ensembles for the Class Imbalance Problem: Bagging-, Boosting-, and Hybrid-Based Approaches. IEEE Transactions on Systems, Man, and Cybernetics, Part C, 463–484 (2012)
3. Guo, et al.: On the Class Imbalance Problem. In: Proceedings of the 2008 Fourth International Conference on Natural Computation, vol. 04, pp. 192–201 (2008)
4. Elkan: The foundations of cost-sensitive learning. In: International Joint Conference on Artificial Intelligence, pp. 973–978 (2001)
5. Garcia, et al.: Exploring the Performance of Resampling Stategies for the Class Imbalance Problem. In: 23rd International Conference on Industrial Engineering and Other Applications of Applied Intelligent Systems, pp. 541–549 (2010)
6. Orriols-Puig, et al.: The class imbalance problem in learning classifier systems: a preliminary study. In: GECOO Workshops, pp. 74–78 (2005)

Web Performance Forecasting with Kriging Method

Leszek Borzemski and Anna Kamińska-Chuchmała

Abstract. Due to the substantial growth of communication network in last years, the access to the Internet network is crucial for the society. Therefore, there is a necessity of research on Web systems forecasting. This work presents a proposal of the application of the geostatistical estimation - the Kriging method, which give spatio-temporal information about forecast of network throughput. The database was created on the basis of Multiagent Internet Measurement System MWING. In the research the connections between an agent in Gdańsk and European servers were considered. The preliminary structural analysis of the data, which are necessary to use the Kriging method was conducted. Next a spatial forecast of the total time of downloading data from Web servers with a four days time advance was calculated. The results were analyzed and compared with other simulation methods results from the same database.

1 Introduction

One of the main reason of increasing traffic in Internet are mobile devices such as smartphones or tablets. Thanks to them, the users usually have the access to the Internet wherever they are and whenever they want. The necessity of good Web performance should be in the focus of the investors and operators of the Web during making decisions about the development of infrastructure. The spatial Web performance forecast could be helpful in this regard, especially during near-term planning of Web of Things. This kind of prediction could help industries to minimize the required expenditure on IT system development.

There exist very little reported research on spatio-temporal web performance forecast. That is the reason way authors examine this issue. The initial work of authors concerns the investigation of two geostatistical simulation methods: Turning

Leszek Borzemski · Anna Kamińska-Chuchmała
Institute of Informatics, Wrocław University of Technology, Wrocław, Poland
e-mail: {leszek.borzemski,anna.kaminska-chuchmala}@pwr.wroc.pl

M. Ali et al. (Eds.): *Contemporary Challenges & Solutions in Applied AI*, SCI 489, pp. 149–154.
DOI: 10.1007/978-3-319-00651-2_20   © Springer International Publishing Switzerland 2013

Bands and Sequential Gaussian Simulation to obtain 3D Web performance forecast, see e.g. [5], [7]. A great advantage of geostatistical methods is the possibility to make area-time forecasts, in which the minimum amount of input information is required, and at the same time the geographical location of Web servers and the total download time of a given resource are taken into account. The natural extension of these research is to utilize one of the geostatistical estimation method, simple kriging, in Web performance forecast, what will be presented in this paper. Knowledge about the throughput forecast and transfer capacity between these web servers and investigated sever will allow to choose the server from which one can receive the required resource in the shortest possible time. These geostatistical methods are using in web systems domain for the first time by the authors. Previously these methods were used mainly in such areas as geology [8], oceanography [9] or economic analysis [1]. Moreover, geostatistical methods have been also used to study the problem of spatial distribution of floating car speed [11] that seems to be similar to the problem of traffic data packets on the Internet. Currently, to the best of the authors knowledge the geospatial approach to Web performance prediction presented in this paper is unique as developed in our papers, leaving no similar problem statement in the literature.

2   Simple Kriging Method

Simple kriging method is a spatial regression named also as kriging with known mean. Kriging method was invented in the early 1950s by Daniel G. Krige which was mining engineer. Simple kriging is used to estimate residuals, where average m is given a priori:

$$Z^*(x_0) = m + \sum_{\alpha=1}^{n} \omega_\alpha (Z(x_\alpha) - m), \tag{1}$$

where:
ω_α - weights attached to the residuals $Z(x_\alpha) - m$,
m - average,
$Z(x_\alpha)$ - a random variable at each of the n locations constructed the data locations x_α.
It is important to mention, that the expected value and the covariance are both translation invariant over the domain D. It means, that for a vector h linking any two points x and $x + h$ in the domain:

$$E[Z(x+h)] = E[Z(x)], \tag{2}$$

$$cov[Z(x+h), Z(x)] = C(h), \tag{3}$$

$$E[Z(x)] = m. \tag{4}$$

Thus, we obtain the expected value from (4), which is the same at any point x of the domain. The covariance between any pair of locations depends only on the vector h which separates them.

The estimation error is the difference between the estimated and the true value at x_0:

$$Z^*(x_0) - Z(x_0). \tag{5}$$

The estimator is unbiased when the estimation error is zero on average:

$$E[Z^*(x_0) - Z(x_0)] = 0. \tag{6}$$

Kriging is preceded by an analysis of the spatial structure of the data. The representation of the average spatial variability is integrated into the estimation procedure in the form of a variogram model. For more information please see [10].

3 Preliminary Data Analysis

The data used in forecasts, were collected during active measurements made by MWING system, the Internet measurement infrastructure developed in our Institute [2], [3]. The database was collected by MWING agent located in Gdańsk, whose main task was to target, by means of HTTP transactions, European Web servers. The database contained the information about a server's geographical location, which the Gdańsk agent targeted, web performance index (Z), which was the total downloading time of rfc1945.txt file and the time stamp of taking a measurement. The data were measured in the interval between 7th and 28th of February 2009 and they were taken every day at the same time: at 6:00 a.m., 12:00 a.m., and 6:00 p.m.

Table 1 Elementary statistics of download times from Web servers between 7-28.02.2009 [6]

Statistical parameters	6:00 a.m.	12:00 a.m.	6:00 p.m.
Minimum value Z_{min} [s]	0.11	0.12	0.12
Maximum value Z_{max} [s]	29.06	12.15	7.93
Average value Z [s]	0.60	0.62	0.60
Standard deviation S [s]	1.59	1.07	0.77
Variability coefficient V [%]	265.00	172.58	128.33
Skewness coefficient G	15.35	7.27	4.99
Kurtosis coefficient K	265.65	64.48	34.61

The table 1 presents the statistics of download times from considered Web servers. From these data, one could determine the large variability of servers' performance. The largest span of data values are presented for 6:00 a.m., where difference between minimum and maximum values is equal to 28.95 seconds. The variability of data for 6:00 p.m. is smaller in comparison to 6:00 a.m. Moreover, for all hours high value of kurtosis and variability coefficient prove the changeability of the examined process. High values of skewness, which is well above 3, indicates big right side asymmetry of performance distribution for whole hours.

4 3D Forecasting of Web Server Performance Using the Simple Kriging Method

The forecast models used to predict the total time of resource download from the Internet have variogram models presented above, which depend on the forecasted hour on day. Other parameters are the same for all three models. Namely, in estimation the moving neighborhood type was adopted where the search ellipsoid was 8.69° for all three directions in the case of web performance. In the estimation the punctual type was used. 3D forecast was calculated with a four day advance, i.e. it encompassed the period between 1st and 4th March 2009. The table 2 presents global statistics of the forecasted values for 6:00 a.m., 12:00 a.m. and 6:00 p.m., respectively.

Table 2 Global statistics of forecasted Web download times with four days time advance, calculated with kriging method

Geostatistical parameter	Minimum value Z_{min} [s]	Maximum value Z_{max} [s]	Average value Z [s]	Variance S^2 $[s]^2$	Standard deviation [s]	Variability coefficient V [%]
Mean forecasted value Z, for 6:00 a.m.	0.12	3.07	0.46	0.03	0.18	39.13
Mean forecasted value Z, for 12:00 a.m.	0.14	1.66	0.48	0.04	0.21	43.75
Mean forecasted value Z, for 6:00 p.m.	0.12	1.61	0.47	0.05	0.23	48.94

The measurements taken for 6:00 p.m. are characterized by the highest variance and standard deviation. However for measurements taken for 12:00 a.m. obtained results have only slightly lower variability coefficient and dispersion between minimum and maximum values. Averaged mean forecasted error *ext post* for all Web server in a four-day forecast for 6:00 a.m. is equal to 30.87%, for 12:00 a.m. is 18.47% and for 6:00 p.m. equals 19.11%. The smallest error of forecast for 12:00 a.m. could be the effect of using Cubic model function for approximation and lower skewness than at 6:00 a.m.

Figure 1 presents on the raster map the final effects of the forecast. This raster map for the 1st day of prognosis (01.03.2009) for 6:00 p.m. presents the download time from the European Web servers. Each cross corresponds to a different given Web server. Geostatistical methods could give information about performance not only for considered servers, but for a whole considered area. On the map in figure 1, a server with the largest download time is located in Kista, Sweden; moreover, this difference is readily visible when compared to other servers. Furthermore, server with the smallest download time is located in Warsaw, Poland. These samples described how varied is Web performance due to different research hours and days.

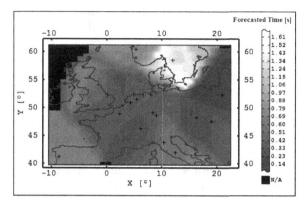

Fig. 1 Sample raster map of download time values from the Internet on 01.03.2009 at 6:00 p.m

5 Comparison Estimation vs. Simulation Methods

Authors made research on the same database and after that compare simulation Turning Bands (TB) method [4], Sequential Gaussian Simulation (SGS) method [6] and presented in this paper Simple Kriging (SK) method.

Table 3 Averaged mean forecasted error *ext post* for all Web servers in a four-day forecast, compared methods

Geostatistical method	6:00 a.m.	12:00 a.m.	6:00 p.m.
Mean forecasted error for SGS	24.83%	16.06%	18.53%
Mean forecasted error for TB	26.91%	20.00%	17.55%
Mean forecasted error for SK	30.87%	18.47%	19.11%

Comparing geostatistical methods, one could conclude, that simulation methods are better. Error of forecast in simulation methods are smaller about 1% to 4%. Estimation method give a worse results especially in cases when in input database the values have the large skewness. It indicate that simulation method further introduce a changeability of examined process. In table 3 there is only one exception where Turning Bands method obtain larger error of forecast than simple kriging (difference about 1.5%). It could be measurement error caused by for example improper selection of variogram model.

6 Summary

In this paper Simple Kriging method was using for Web performance forecasting. This is an innovative approach in considered research area. Large-scale measurement experiment MWING was performed in a real-life Internet to gather the data characterizing performance of many Web servers localized in Europe and perceived from agent installed in Gdańsk. Accuracy of forecasted Web performance described in this paper is slightly below 20% (except special cases with very large value of skewness coefficient), what can be considered as meaningful contribution in this field of research. Authors compared geostatistics methods: simulation and estimation for advantage simulation methods. Estimation methods are slightly less accurate for characterized high variability and changeability of download times values. Nevertheless, we claim that presented estimation method: simple kriging to performance forecasting could be helpful in spatial analysis of Internet and Web performance within a given geographic area.

As future research is planning a new active measurement experiment to gain new data for analysis and making model of spatial forecast.

References

1. Amiri, A., Gerdtham, U.: Relationship between exports, imports, and economic growth in france: evidence from cointegration analysis and granger causality with using geostatistical models. Tech. rep., Munich Personal RePEc Archive Paper No. 34190 (2011)
2. Borzemski, L.: The experimental design for data mining to discover web performance issues in a wide area network. Cybernetics and Systems 41, 31–45 (2010)
3. Borzemski, L., Cichocki, Ł., Fraś, M., Kliber, M., Nowak, Z.: MWING: A multiagent system for web site measurements. In: Nguyen, N.T., Grzech, A., Howlett, R.J., Jain, L.C. (eds.) KES-AMSTA 2007. LNCS (LNAI), vol. 4496, pp. 278–287. Springer, Heidelberg (2007)
4. Borzemski, L., Danielak, M., Kaminska-Chuchmala, A.: Short-term spatio-temporal forecasts of web performance by means of turning bands method. In: Nguyen, N.-T., Hoang, K., Jędrzejowicz, P. (eds.) ICCCI 2012, Part II. LNCS (LNAI), vol. 7654, pp. 132–141. Springer, Heidelberg (2012)
5. Borzemski, L., Kamińska-Chuchmała, A.: Client-perceived web performance knowledge discovery through turning bands method. Cybernetics and Systems 43, 354–368 (2012)
6. Borzemski, L., Kamińska-Chuchmała, A.: Knowledge engineering relating to spatial web performance forecasting with sequential gaussian simulation method. In: Advances in Knowledge-Based and Intelligent Information and Engineering Systems, Frontiers in Artificial Intelligence and Applications, vol. 243, pp. 1439–1448. IOS Press, Amsterdam (2012)
7. Borzemski, L., Kamińska-Chuchmała, A.: Distributed web systems performance forecasting using turning bands method. IEEE Transactions on Industrial Informatics 9, 254–261 (2013)
8. Dubrule, O.: Geostatistics for Seismic Data Integration in Earth Models. Society of Exploration Geophysicists and European Association of Geoscientists and Engineers (2003)
9. Inizan, M.: Geostatistical validation of a marine ecosystem model using in situ data. Tech. rep., Centre de Geostatistique, Ecole des Mines de Paris (2002)
10. Wackernagel, H.: Multivariate Geostatistics: an Introduction with Applications. Springer, Berlin (2003)
11. Wang, Y., Zhuang, D., Liu, H.: Spatial distribution of floating car speed. Journal of Transportation Systems Engineering and Information Technology 12, 36–41 (2012)

Part VIII
Problem Solving

Application of the Swarm Intelligence Algorithm for Investigating the Inverse Continuous Casting Problem

Edyta Hetmaniok, Damian Słota, and Adam Zielonka

Abstract. In the paper a proposal of procedure for solving the inverse problem of continuous casting is presented. The proposed approach consists in applying the swarm intelligence algorithm imitating the behavior of ants for minimizing an appropriate functional which enables to determine the unknown cooling conditions of the process.

1 Introduction

Many problems of technical or engineering nature lead to optimization tasks. In recent times a great popularity in solving optimization problems have been found by the algorithms of artificial intelligence, a specific group of which is represented by the swarm intelligence (SI) algorithms based on the collective behavior of the swarm members [4]. Particular versions of the swarm intelligence algorithms differ in the ways of communication between individuals and exploration of considered space. For example, basic idea of the Ant Colony Optimization (ACO) algorithm is inspired by the behavior of real ants exploring the environment in order to find the best path leading to the source of food [3]. During this search they lay down a chemical substance, called the pheromone, directing each other ant to the best path. Such behavior is imitated by the artificial "ants" condensing their presence around the best located solutions. Inventors of artificial intelligence algorithms were looking for inspirations mostly in world or animals or other living organisms [2, 7]. However, there have also appeared some algorithms inspired by the human behavior. Example of such approach is the Harmony Search algorithm (HS) [5] based on the similarity between the process of jazz improvisation and the problem of function optimizing.

Edyta Hetmaniok · Damian Słota · Adam Zielonka
Institute of Mathematics, Silesian University of Technology, Kaszubska 23, Gliwice, Poland
e-mail: {Edyta.Hetmaniok,Damian.Slota,Adam.Zielonka}@polsl.pl

M. Ali et al. (Eds.): *Contemporary Challenges & Solutions in Applied AI*, SCI 489, pp. 157–162.
DOI: 10.1007/978-3-319-00651-2_21   © Springer International Publishing Switzerland 2013

The inverse problems for the equations of mathematical physics consist in determination of some missing elements which can be, for example, the initial condition, boundary conditions or parameters of material, and the missing part of input information is compensated by the additional information about the consequences resulting from the input conditions [6, 1]. Difficulty in solving such problems results from the fact that the analytic solution may not exist or it exists but is neither unique nor stable. Problem considered in this paper is the inverse problem of continuous casting, it means of the process whereby the molten metal is poured in the controlled way into the crystallizer where it solidifies by taking the appropriate form and then is consecutively moved out from there [8, 9, 10].

2   Two-Dimensional Inverse Continuous Casting Problem

Let us consider the continuous casting of pure metals on a vertical device working in the undisturbed cycle. Under appropriate conditions (see [9, 10]) and because of the heat symmetry, the region Ω of created ingot can be considered as the two-dimensional region consisted of two subregions: Ω_1 taken by the liquid phase and Ω_2 taken by the solid phase, separated by the freezing front Γ_g (described by means of function $x = \xi(t)$). In these subregions the heat transfer process, including the apparently steady field of temperature and location of the freezing front, can be described by the two-phase Stefan problem [8]. Boundary of region $\Omega = [0,b] \times [0,z^*] \subset \mathbb{R}^2$ is divided into four subsets where the boundary conditions are defined: $\Gamma_0 = \{(x,0); \ x \in [0,b]\}$, $\Gamma_1 = \{(0,z); \ z \in [0,z^*]\}$, $\Gamma_2 = \{(b,z); \ z \in [0,z_1]\}$ and $\Gamma_3 = \{(b,z); \ z \in (z_1,z^*]\}$.

Discussed problem consists in determination of the cooling conditions for the ingot in such way that the temperature in selected points of the solid phase would take the given values ($(x_i,z_j) \in \Omega_2$):

$$T_2(x_i,z_j) = U_{ij}, \qquad i = 1,2,\ldots,N_1, \quad j = 1,2,\ldots,N_2, \tag{1}$$

where N_1 denotes the number of sensors and N_2 describes the number of measurements taken from each sensor. Another elements which should be determined are: the function ξ describing the freezing front location and the functions of temperature T_k in regions Ω_k ($k = 1,2$). Functions of temperature within the regions Ω_k (for $k = 1,2$) satisfy the heat conduction equation

$$c_k \rho_k w \frac{\partial T_k}{\partial z}(x,z) = \frac{\partial}{\partial x}\left(\lambda_k \frac{\partial T_k}{\partial x}(x,z)\right), \tag{2}$$

where c_k, ρ_k and λ_k denote, respectively, the specific heat, mass density and thermal conductivity in liquid phase ($k = 1$) and solid phase ($k = 2$), w is the velocity of continuous casting, and, lastly, x and z denote the spatial variables.

On the respective parts of boundary the appropriate boundary conditions must be satisfied:

$$T_1(x,0) = T_z, \quad (T_z > T^*) \quad \text{on } \Gamma_0, \tag{3}$$

$$\frac{\partial T_k}{\partial x}(x,z) = 0 \quad \text{on } \Gamma_1, \tag{4}$$

$$-\lambda_k \frac{\partial T_k}{\partial x}(x,z) = q(z) \quad \text{on } \Gamma_2, \tag{5}$$

$$-\lambda_k \frac{\partial T_k}{\partial x}(x,z) = \alpha(z)\left(T_k(x,z) - T_\infty\right) \quad \text{on } \Gamma_3, \tag{6}$$

$$T_1\left(\xi(z),z\right) = T_2\left(\xi(z),z\right) = T^* \quad \text{on } \Gamma_g, \tag{7}$$

$$L\rho_2 w \frac{d\xi(z)}{dz} = -\lambda_1 \left.\frac{\partial T_1(x,z)}{\partial x}\right|_{x=\xi(z)} + \lambda_2 \left.\frac{\partial T_2(x,z)}{\partial x}\right|_{x=\xi(z)} \quad \text{on } \Gamma_g, \tag{8}$$

where α describes the heat transfer coefficient, q denotes the heat flux density, T_z is the pouring temperature, T_∞ is the ambient temperature, T^* is the solidification temperature and L describes the latent heat of fusion.

In considered approach the sought elements are the heat flux and the heat transfer coefficient, it means the following function f should be determined

$$f(z) = \begin{cases} q(z) & \text{for } z \leq z_1, \\ \alpha(z) & \text{for } z > z_1. \end{cases} \tag{9}$$

For the fixed form of function f problem (2)–(8) turns into the direct Stefan problem, solving of which enables to find the courses of temperature $T_{ij} = T_2(x_i,z_j)$ corresponding to function f. By using the calculated temperatures T_{ij} and the given temperatures U_{ij} the following functional is constructed

$$J(f) = \sum_{i=1}^{N_1} \sum_{j=1}^{N_2} \left(T_{ij} - U_{ij}\right)^2, \tag{10}$$

representing the error of approximate solution. Since our goal is to find such form of function f that the reconstructed temperatures will be as close as possible to its measurement values, solving of considered problem reduces to minimization of functional (10) by applying the ACO algorithm. For solving the direct Stefan problem the alternating phase truncation method [10] is used.

3 Swarm Intelligence Algorithm

Imitation of ants' behavior in artificial space is the following. Let $F(\mathbf{x})$, for $\mathbf{x} \in D$, be the minimized function. The procedure is initialized by random selection of the initial ants localizations, it means vectors $\mathbf{x}^k = (x_1^k, \ldots, x_n^k)$, where $\mathbf{x}^k \in D$, $k = 1,2,\ldots,m$, from among which the best located ant $\mathbf{x}^{best}$ (such that the minimized function takes the lowest value) is determined.

In every iteration to each vector $\mathbf{x}^k$ a modification based on the pheromone trail is applied according to the following formula: $\mathbf{x}^k = \mathbf{x}^{best} + \mathbf{dx}^k$, where $\mathbf{dx}^k$ is a vector determining the length of jump. Elements of this vector are randomly generated from the interval $[-\beta, \beta]$ (where $\beta = \beta_0$ is the narrowing parameter defined in the initialization of the algorithm). The best located ant $\mathbf{x}^{best}$ in the current ants' population is next determined. These two steps are repeated I^2 times, where I denotes the assumed number of iterations. At the end of every iteration the range of ants dislocations is decreasing, according to the formula $\beta_{i+1} = 0.1\beta_i$, which simulates the evaporation of pheromone trail in nature. More information about the construction of ACO algorithm can be found in [3].

4 Numerical Example

As the example let us consider the continuous casting of aluminium described by the following parameters [10]: $b = 0.1$ [m], $\lambda_1 = 104$ [W/(m K)], $\lambda_2 = 204$ [W/(m K)], $c_1 = 1290$ [J/(kg K)], $c_2 = 1000$ [J/(kg K)], $\rho_1 = 2380$ [kg/m^3], $\rho_2 = 2679$ [kg/m^3], $L = 390000$ [J/kg], velocity of casting $w = 0.002$ [m/s], solidification temperature $T^* = 930$ [K], ambient temperature $T_\infty = 298$ [K] and pouring temperature $T_z = 1013$ [K]. Known cooling conditions, which we intend to reconstruct, are the following [10]: $q(z) = 400000$ [W/m^2] and $\alpha(z) = 4000$ [W/(m^2 K)]. In considered region there are installed two thermocouples ($N_1 = 2$) located 0.001 and 0.002 m away from boundary of the region. From each thermocouple we used 100 measurements of temperature ($N_1 = 100$). Distance between the successive measurements is equal to 0.002 m. In calculations we used the exact values of temperature and values burdened by the random error of normal distribution and values 2 and 5%.

ACO algorithm was executed for number of ants $m = 15$ and number of iterations $I = 6$. Initial value of narrowing parameter was different for each of identified parameters because of the big difference between expected values of reconstructed parameters ($\beta_0 = 150000$ and $\beta_0 = 2500$ for q and α, respectively). For the same reason the initial ants populations were randomly selected from different intervals ($[250000, 500000]$ and $[1000, 5000]$ for q and α, respectively). Since the ACO algorithm belongs to the group of heuristic algorithms, which means that each execution of the procedure can give slightly different results, we evaluated the calculations for 20 times and the best of received results were taken as the reconstructed elements.

In Figure 1 the relative errors of the heat flux q and the heat transfer coefficient α identification in dependence on the number of iterations obtained for input data burdened by 2% and 5% error, respectively, is displayed. We can see that in both cases we obtained very quickly, in almost two iterations, very good reconstructions and the further iterations did not improve significantly the results. For the unburdened input data the reconstruction errors converge very quickly to zero which confirms the stability of used procedure.

Figure 2 presents the reconstructed distributions of temperature in measurement points located 0.002 m away from boundary of the region, compared with the known exact distributions. The results are obtained for input data burdened by 5% error and one can observe that even in this worst of considered cases the reconstructed and known courses of temperature are almost the same. The absolute error of this

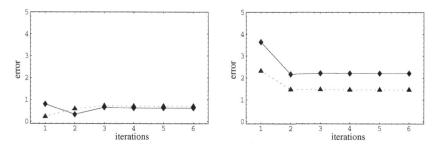

Fig. 1 Relative errors of parameter f reconstruction for the successive iterations (▲ – for q, ♦ – for α) obtained for 2% (left figure) and 5% (right figure) noise of input data

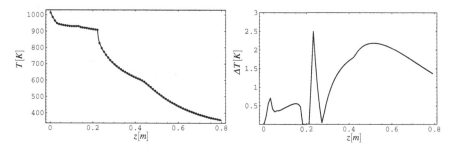

Fig. 2 Exact (solid line) and reconstructed (dots) distributions of temperature (left figure) in control point located 0.002 m away from boundary b obtained for 5% noise of input data and absolute error of this reconstruction (right figure)

Table 1 Reconstructed values of f, relative errors (δ_f), standard deviations (σ), standard deviations (σ^p) expressed as a percent of mean values of f and maximal relative errors (δ_T^{max}) of temperature reconstruction obtained for various noises of input data

noise	f	δ_f [%]	σ	σ^p [%]	δ_T^{max}[%]
0%	400380.19	0.09506	380.254	0.09506	0.00002
	4001.56	0.03902	1.572	0.03930	
2%	397543.07	0.61422	2456.931	0.61423	0.00254
	4028.20	0.70513	28.205	0.70512	
5%	408832.51	2.20814	8832.541	2.20814	0.00461
	3941.62	1.45966	58.386	1.45964	

reconstruction is at the level of few kelvins which additionally confirms the almost perfect reconstruction of temperature.

Finally, in Table 1 the statistical elaboration of results obtained in 20 executions of the procedure for various noises of input data is compiled. Relative errors of the heat flux and the heat transfer coefficient reconstruction for unburdened input data are smaller than 0.1%. The errors increase obviously with the increasing value of input data perturbation but still in each case is much smaller than the input data error. Stability of the procedure is indicated by the standard deviations of the sought parameters reconstruction which in most cases are the parts of percent of the mean values of these reconstructions.

5 Conclusions

The paper contains a description of the procedure for solving the inverse problem of continuous casting. Presented results indicate that the elaborated method ensures the approximate solution rapidly convergent to exact solution and perturbated by the error not exceeding the error of input data for relatively small number of iterations. Therefore the advantages of using approach with the ACO algorithm, in comparison with classical methods, are the short time of working and no particular assumptions needed for minimized functional.

Acknowledgements. This project has been financed from the funds of the National Science Centre granted on the basis of decision DEC-2011/03/B/ST8/06004.

References

1. Beck, J.V., Blackwell, B., St. Clair, C.R.: Inverse Heat Conduction: Ill Posed Problems. Wiley Intersc., New York (1985)
2. Chu, S.-C., Tsai, P.-W., Pan, J.-S.: Cat Swarm Optimization. In: Yang, Q., Webb, G. (eds.) PRICAI 2006. LNCS (LNAI), vol. 4099, pp. 854–858. Springer, Heidelberg (2006)
3. Duran Toksari, M.: Ant Colony Optimization for finding the global minimum. Appl. Math. Comput. 176, 308–316 (2006)
4. Eberhart, R.C., Shi, Y., Kennedy, J.: Swarm Intelligence. Morgan Kaufmann, San Francisco (2001)
5. Geem, Z.W.: Improved Harmony Search from ensemble of music players. In: Gabrys, B., Howlett, R.J., Jain, L.C. (eds.) KES 2006. LNCS (LNAI), vol. 4251, pp. 86–93. Springer, Heidelberg (2006)
6. Hetmaniok, E., Słota, D., Zielonka, A., Wituła, R.: Comparison of ABC and ACO algorithms applied for solving the inverse heat conduction problem. In: Rutkowski, L., Korytkowski, M., Scherer, R., Tadeusiewicz, R., Zadeh, L.A., Zurada, J.M. (eds.) EC 2012 and SIDE 2012. LNCS, vol. 7269, pp. 249–257. Springer, Heidelberg (2012)
7. Mehrabian, R., Lucas, C.: A novel numerical optimization algorithm inspired from weed colonization. Ecological Informatics 1(4), 355–366 (2006)
8. Nowak, I., Nowak, A.J., Wrobel, L.C.: Inverse analysis of continuous casting processes. Int. J. Numer. Methods Heat Fluid Flow 13, 547–564 (2003)
9. Słota, D.: Identification of the cooling condition in 2-D and 3-D continuous casting processes. Numer. Heat Transfer B 55, 155–176 (2009)
10. Słota, D.: Solving of the inverse solidification problems by using the genetic algorithms. Silesian University of Technology Press, Gliwice (2011) (in Polish)

Estimating Mental States of a Depressed Person with Bayesian Networks

Michel C. A. Klein and Gabriele Modena

Abstract. In this *work in progress* paper we present an approach based on Bayesian Networks to model the relationship between mental states and empirical observations in a depressed person. We encode relationships and domain expertise as a Hierarchical Bayesian Network. Mental states are represented as latent (hidden) variables and the measurements found in the data are encoded as a probability distribution generated by such latent variables; we provide examples of how the network can be used to estimate mental states.

1 Introduction

Major depression is currently the fourth disorder worldwide in terms of disease burden, and is expected to be the disorder with the highest disease burden in high-income countries by 2030. Within the context of the FP7 project ICT4Depression an intelligent support system has been developed to assist patients suffering from depression. Mobile devices are used for monitoring activities and biosignals in a non-intrusive and continuous way. A model-based approach is used to estimate the current influence and future developments of a therapy on a patient. By means of computational simulation of states associated with depression and therapeutic models [3], [6], [2], [5] (a so-called virtual patient [4]), tailored feedback and therapeutic advice is provided to the patient. States considered in the simulation are *mood*, *openness to therapy*, *coping skills* and *vulnerability*. Although we cannot directly observe mental states, a lot of patient data is collected by the system. Such data includes physiological measurements, activity logs, journal, social signals, questionnaire, mood ratings [10]. In this paper we introduce a *work in progress* approach to link mental states to data. We propose a method based on causal networks and

Michel C.A. Klein · Gabriele Modena
VU University Amsterdam
e-mail: gabriele.modena@gmail.com, michel.klein@cs.vu.nl

M. Ali et al. (Eds.): *Contemporary Challenges & Solutions in Applied AI*, SCI 489, pp. 163–168.
DOI: 10.1007/978-3-319-00651-2_22   © Springer International Publishing Switzerland 2013

Bayesian inference to model relationships between mental states and empirical observations. There are at least two different use cases for such a model. The first — straightforward— use case is calculating the probability for a specific mental state given certain observations. Secondly, the model can be used to infer mental states from empirical measurement data. This mechanism is exploited for estimating parameters in a personalized system that diagnoses a patient and provides support. In a later stage the method will be evaluated and further developed using data collected in two trials with actual patients scheduled to start in the third quarter of 2012 and first quarter of 2013. These trials will be carried out in two *out of a lab settings* and will provide a source of unique data.

2 Method

Bayesian Networks (BN) are a modelling technique that uses conditional independence assumptions to perform calculations on local sets of variables. These lead to results on the whole distribution of variables. Modelling using BNs consists of two phases. First there is a hierarchical organization of variables identified on the domain using a graph that describes the structure of a probabilistic model. Then, for each variable, the characterization of its probability distribution, as a conditional or a marginal distribution according to the topology of the graph. The set of distributions allows to recover the joint distribution on the domain [1]. Because a Bayesian Network is a complete model for the variables and their relationships, it can be used to answer probabilistic queries about them. For example, the network can be used to find out updated knowledge of the state of a subset of variables when other variables (the evidence variables) are observed. The concept of conditional independence for two variables X and Y given a third variable Z means that the stochastic variation of X and Y becomes independent once the variable Z is known. Bayesian network modelling intensively makes use of this property. It allows modelling dependencies on reduced sets of variables. A Bayesian network can thus be considered a mechanism for automatically applying Bayes' theorem to complex problems.

To our knowledge Bayesian Networks have not been widely used in the context of E-Mental Health. In this paper we follow an approach similar to [7]. We encode domain expertise in a Bayesian Network model. We represent mental states as latent (hidden) variables and the measurements found in the data as a probability distribution generated by such latent variables. In contrast to the approach in [7] we are interested in modelling cognitive aspects of a patient, rather than neurobiological ones. Moreover we consider the case in which the measurements are influenced by lower level records present in the data. The methodology is oriented to the discovery of a link between the latent variable Θ and each measurement M_i. An example of a Bayesian Network to model this scenario is given in Figure 1 (left). The figure depicts three measurements M_1, M_2, M_3 influenced by a latent variable Θ representing a mental state.

Figure 1 (right) shows a second type of relationship between nodes. In this topology we consider measurements being influenced by lower level signals in the data

Fig. 1 Left: latent variables and measurements (standard Bayesian network topology). Right: example of a Hierarchical Bayesian Network.

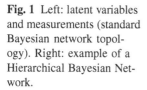
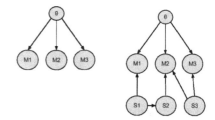

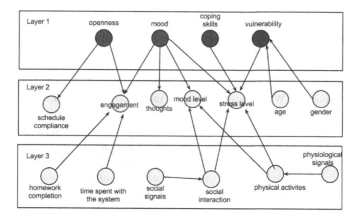

Fig. 2 Example of modelling relationships between mental states (dark nodes) and measurements (light nodes) in with a hierarchical Bayesian Network.

S_1, S_2, S_3. This topology is usually referred to in literature as Hierarchical Bayesian Network (HBN) [8]. Intuitively HBNs can be considered as a generalization of standard Bayesian Networks where a node can be an aggregate data type. HBNs encode conditional independency the same way as standard BNs but allow to express further knowledge about variable structure. HBNs can be exploited to build more realistic probabilistic models.

For the purpose of this paper we model attributes of the network (nodes) as discrete random variables. Numerical and categorical measurements, such as age, gender and mood level, are discretized and the network attributes are represented as Conditional Probability Distribution (CPD) tables. As a first step we construct a three layer Hierarchical Bayesian Network that encodes relationships between mental states and measurements recorded in data. The network is designed based on literature and feedback provided by domain experts. Figure 2 depicts the network. The first layer contains the latent variables of the model that represent mental states. These are openness, mood, coping skills and vulnerability of a person. Mental states influence the variables in the second layer. These nodes represent measurements collected by the system as aggregate data types. Finally a third layer models explicitly how such aggregates are influenced by lower level records present in the data.

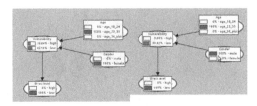

Fig. 3 Querying the network

3 Examples

Querying the Network

First, we give an example of using the resulting Bayesian Network for inferencing probabilities. We consider the case in which *vulnerability* can have two states; high and low. We want to model the property that *high vulnerability is more likely to cause a high level of stress in female depressed subjects in the age range 25-35 than in male subjects from the same demographics.* We encode the relationship between vulnerability, age and gender of a patient and stress level using the SamIam tool[1]. In this example we assume that genders and age groups are evenly distributed. The probability distribution of the *vulnerability* variable is then factored as shown in Table 1. The probabilities for the *stress* variable are as follows: a high vulnerability gives a 0.8 chance on high stress, while a low vulnerability is associated with a probability of 0.2 on high stress.

Figure 3 shows how the network can be queried to estimate the vulnerability state (marked in green) given evidence about stress level, age group and gender (marked in red). To do so we set the vulnerability node as hidden on the en-

Table 1 Patient vulnerability given gender and age

Age	18-25		25-35		Over 35	
Gender	male	female	male	female	male	female
high	0.6	0.4	0.3	0.7	0.45	0.55
low	0.4	0.6	0.7	0.3	0.55	0.45

coded network and we fill in evidence in the model for a specific case. As example, we do this for two different cases: 1) a male patient in the age group 25-35 with a low stress level, and 2) a female patient with similar characteristics. When we query the network to infer the probability of the vulnerability state being high or low for both scenarios, we find the following outcome (see also Figure 3):
$P(Vulnerability = high|stress\_level = low, gender = M, age = 25 - 35) = 36.85\%$
whereas $P(Vulnerability = high|stress\_level = low, gender = F, age = 25 - 35) = 9.68\%$. These values are consistent with the property we wanted to model.

Parameter Estimation

The second, more interesting, example is the usage of the model for estimating parameters of a system that provides personalized guidance for a patient. For this example, we assume that the system uses the mental states 'mood', 'openness',

---

[1] http://reasoning.cs.ucla.edu/samiam/

'coping skills' and 'vulnerability' as hidden variables. The system collects data about the behaviour of the patient (e.g. activities performed, time spent with the system, etc.) and should use this data to estimate the aforementioned variables. A preliminary experiment using artificially generated data showed promising results in terms of consistency and robustness of the network. We carried out the experiment as follows. First we created three patient profiles and encoded their states in two topologies of Bayesian network. The profiles are shown in Table 2. A patient profile is characterized by a combination of mental states. We consider the cases in which a patient has high mood and coping skill levels and low openness and vulnerability; a second patient has high vulnerability but low states for all other aspects. Finally a third case represents a patient showing high mood, openness and coping skill while at the same time having low vulnerability.

Then, we considered two different Bayesian Networks: the Hierarchical Bayesian Network depicted in Figure 2 and a standard Bayesian Network obtained by removing the third layer of nodes. We encoded both networks using the Bayes Net Toolbox for Matlab [2] software. For both networks we assumed that all variables are observed and for each variable we manually initialized states priors. In both networks we first initialize mental states according to the profiles depicted in Table 2. This leads to three distinct probability factorizations for each network topology. We artificially created a dataset D that mimics measurements generated by the three patient profiles in both networks. Recalling that each measurement is represented as a discrete random variable m, D is a set of vectors in the form $(m_1 = s_i, ..., m_N = s_j)$ for each m and each discrete state s associated to that random variable.

In the third step of the experiment we marked mental states as hidden variables and the lower layers of the network as observed. Given a network topology, a factored probability distribution on the known variables and an artificially generated dataset of possible discrete states configuration we want to estimate the original configuration of mental states (hidden variables). That is, for each hidden variable Θ we want to

Table 2 Patient profiles. Each line represents a patient

mood	openness to therapy	coping skills	vulner- ability
high	low	high	low
low	low	low	high
high	high	high	low

compute $P(\Theta|D)$. We did so by attempting both Maximum-Likelihood Estimation and Bayesian Estimation with Dirichlet priors [9]. This mimics the situation in which we do have measurements about the lower layer concepts, but need to estimate the parameters in the hidden layer. In all these cases we have been able to correctly identify all latent variable configurations that generated the probability distribution factored in both network topologies. This illustrates how a Bayesian Network that is instantiated with the right probability distribution in principle can be used to infer mental states based on a set of measurements. Although our preliminary experiments suggest consistency of the chosen topologies, at this stage it is not possible to determine which one is a better model. Further experiments with actual data are therefore necessary to understand how to better model this domain.

---

[2] https://code.google.com/p/bnt/

The same consideration applies to parameter learning. Although in this preliminary experiment we have been able to correctly estimate hidden variables, it is not yet clear which approaches can lead to better results.

4 Future Work

In this preliminary work we presented a method based on the methodology of Bayesian Networks to estimate mental states in depressed persons from data collected by an Intelligent Support System. Probabilistic Graphical Models are widely used in domains that span from robotics to medicine. To our knowledge, however, the paradigm has not yet been widely explored within the E-Mental Health domain. Due to a general lack of mental health datasets in the public domain we reported experiments performed on artificially generated data. In the continuation of this work we will address technical issues related to parameter estimation, network encoding and simulation of the models and contrast it to other methods using a dataset collected during two *out of a lab* medical trials.

References

1. Ben-Gal, I.: Bayesian Networks. John Wiley and Sons (2007)
2. Both, F., Cuijpers, P., Hoogendoorn, M., Klein, M.C.A.: Towards fully automated psychotherapy for adults - bas - behavioral activation scheduling via web and mobile phone. In: HEALTHINF, pp. 375–380 (2010)
3. Both, F., Hoogendoorn, M.: Utilization of a virtual patient model to enable tailored therapy for depressed patients. In: Lu, B.-L., Zhang, L., Kwok, J. (eds.) ICONIP 2011, Part III. LNCS, vol. 7064, pp. 700–710. Springer, Heidelberg (2011)
4. Both, F., Hoogendoorn, M., Klein, M., Treur, J.: Modeling the dynamics of mood and depression. In: Proceedings of the 2008 conference on ECAI 2008: 18th European Conference on Artificial Intelligence, pp. 266–270. IOS Press, Amsterdam (2008)
5. Both, F., Hoogendoorn, M., Klein, M.C.A., Treur, J.: Computational modeling and analysis of the role of physical activity in mood regulation and depression. In: Wong, K.W., Mendis, B.S.U., Bouzerdoum, A. (eds.) ICONIP 2010, Part I. LNCS, vol. 6443, pp. 270–281. Springer, Heidelberg (2010)
6. Both, F., Hoogendoorn, M., Klein, M.C.A., Treur, J.: Computational modeling and analysis of therapeutical interventions for depression. In: Yao, Y., Sun, R., Poggio, T., Liu, J., Zhong, N., Huang, J. (eds.) BI 2010. LNCS, vol. 6334, pp. 274–287. Springer, Heidelberg (2010)
7. Chevrolat, J.-P., Golmard, J.-L., Ammar, S., Jouvent, R., Boisvieux, J.-F.: Modelling behavioral syndromes using bayesian networks. Artificial Intelligence in Medicine 14(3), 259–277 (1998)
8. Gyftodimos, E., Flach, P.: Hierarchical bayesian networks: A probabilistic reasoning model for structured domains. In: Proceedings of ICML Workshop on Development of Representations (2002)
9. Heckerman, D.: A tutorial on learning with bayesian networks (1995)
10. Rocha, A., Henriques, M., Lopes, J., Camacho, R., Klein, M., Modena, G., de Ven, P., McGovern, E., Tousset, E., Gauthier, T., et al.: Ict4depression: Service oriented architecture applied to the treatment of depression. In: 2012 25th International Symposium on Computer-Based Medical Systems (CBMS), pp. 1–6. IEEE (2012)

Multi-objective Optimization Algorithms for Microchannel Heat Sink Design

Ahmed Mohammed Adham, Normah Mohd-Ghazali, and Robiah Ahmad[*]

Abstract. This paper investigates the performance of four multi-objective optimization algorithms namely the GAM, MOGA, SPEA2 and NSGA-II on the optimization of a microchannel heat sink based on the total thermal resistance and pumping power. Two case studies with different formulation methodologies were selected for the optimizations. The optimizations results showed that both SPEA2 and NSGA-II algorithms exhibited excellent performance in terms of the number of the optimal solutions, maintaining the desirable diversity and convergence speed toward the Pareto optimal front as compared to GAM and MOGA.

Keywords: Multi-objective, Algorithms, Optimization, Heat Sink.

1   Introduction

The main issue today with powerful electronic chips is that they produce a huge quantity of heat. Microchannel heat sink (MCHS) is one of the methods used to remove this heat. A typical MCHS consists of a substrate (t) from a high thermal conductivity material with many microchannels attached on top, as shown in Fig. 1. Initial optimization of the performance focused on the microchannel geometries, structural materials, and coolants [1]. Later, optimization techniques with numerical schemes, analytical methods, and recently genetic algorithm and

Ahmed Mohammed Adham · Normah Mohd-Ghazali
Faculty of Mechanical Engineering, Universiti Teknologi Malaysia
81310 Skudai, Johor Bahru, Malaysia

Robiah Ahmad
UTM Razak School of Engineering and Advanced Technology, UTM KualaLumpur
54100 Jalan Semarak, Kuala Lumpur, Malaysia
e-mail: robiah@ic.utm.my

[*] Corresponding author.

M. Ali et al. (Eds.): *Contemporary Challenges & Solutions in Applied AI*, SCI 489, pp. 169–174.
DOI: 10.1007/978-3-319-00651-2_23   © Springer International Publishing Switzerland 2013

evolutionary approaches [2] were developed to solve complex mathematical functions efficiently with single and multi-objective problems [3]. A hybrid technique using the elitist non-dominated sorting genetic algorithm (NSGA-II) and the sequential quadratic programming (SQP) were also used as the main and local search tool for the optimization of the MCHS. This study investigates the capabilities of some other multi-objective optimization algorithms in the area of MCHS optimization such as the goal attainment method (GAM) [4], multi-objective genetic algorithm (MOGA) [3], strength Pareto evolutionary algorithm (SPEA2) [5] and elitist non-dominated sorting genetic algorithm (NSGA-II) [6].

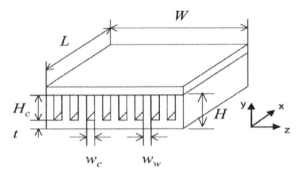

Fig. 1 Physical model of microchannel heat sink used for electronic cooling

2 Multi-objective Optimization Algorithms

GAM, proposed by Gembicki [4], converts the multi-objective functions into a single-objective function system by inserting appropriate weight factors for each objective. It is expressed as,

$$\min_{x \in \Omega} \lambda \text{ Such that } \quad f_i - w_i \lambda \leq g_i$$

where w_i are the weight factors and g_i the goals to be attained by the objective functions f_i, estimated by the decision maker prior to the optimization process. Minimizing λ (slack variable) produces a non-dominated solution which either over or under predicts the specified goal.

MOGA employed the non-dominated sorting concept [3] if a wide spread of solutions within the feasible objective space is desired. SPEA2 which was proposed by Zitzler and Thiele [6], is as an improved version of its predecessor, the SPEA.

Deb et al. [7] proposed NSGA-II to overcome the poor distribution of the optimal solutions along the Pareto optimal front and high computing complexity of the NSGA.

3 Test Functions

Two case studies, case study 1 and case study 2, are selected from the MCHS optimization field with the thermal resistance and pumping power as objective test functions. These functions were minimized using four multi-objective algorithms (GAM, MOGA, SPEA2 and NSGA-II) with the population size of 4, number of generation of 100 and selection function used is tournament. The objective functions for case 1, was developed by Husain and Kim [2]:

$$R_{th} = 0.0964 + 0.3124\theta - 0.7005\varphi - 1.1122\theta\varphi + 0.6044\theta^2 + 4.8528\varphi^2 \tag{1}$$

$$\overline{P} = 0.9925 - 9.3955\,\theta + 3.5575\,\phi - 14.9250\,\theta\phi + 22.9024\,\theta^2 - 0.0706\,\phi^2 \tag{2}$$

where R_{th} and $\overline{P}$ are the thermal resistance and pumping power, respectively. They consist of two design variables, θ and ϕ, which are related to the rectangular microchannel geometry. The limits of these variables are set as $0.1 \leq \alpha \leq 0.25$ and $0.04 \leq \beta\,0.1$ [2]. The objective functions in case 2 are [6],

$$R_{total} = \frac{L}{Cp_f\mu_f}\frac{2}{Re}\frac{1+\beta}{1+\alpha} + \frac{1}{h_{av}}\frac{1+\beta}{1+2\alpha\eta} + \frac{t}{k_{hs}} + \frac{1+\beta}{\pi k_{hs}}\ln\left[\frac{1}{\sin\dfrac{\pi\beta}{2(1+\beta)}}\right]\frac{H_c}{\alpha} \tag{3}$$

$$P_{ptot} = \Delta p_{tot}.G \tag{4}$$

where Δp_{tot} is the total pressure drop given as,

$$\Delta p_{tot} = f_{hs}\frac{(1+\alpha)L}{2H_c}\rho_f\frac{V_{mf}^2}{2} + (1.79 - 2.23(\frac{1}{1+\beta}) + 0.53(\frac{1}{1+\beta})^2)\rho_f\frac{V_{mf}^2}{2} + f_{t1}\frac{L_{t1}}{D_{tu}}\rho_f\frac{V_{mt}^2}{2}$$
$$+ f_{t2}\frac{L_{t2}}{D_{tu}}\rho_f\frac{V_{mt}^2}{2} + 0.42\rho_f\frac{V_{mt}^2}{2} + (1 - \frac{A_t^2}{A_{hs}^2})\rho_f\frac{V_{mt}^2}{2} + 0.42(1 - \frac{A_t^2}{A_{hs}^2})\rho_f\frac{V_{mt}^2}{2} + \rho_f\frac{V_{mt}^2}{2}$$

4   Results and Discussion

Results are benchmarked against NSGA-II as well as compared against each other in terms of the number of optimal solutions, convergence rate and diversity.

For case study 1, Figs 2(a), (b), (c) and (d) show the performances of the NSGA-II compared to GAM, MOGA and SPEA2, respectively.

It can be seen that NSGA-II outperforms GAM. In a single run, NSGA-II provides ~ 44 optimal solutions while GAM provides only one which converges at a faster rate. NSGA-II gave a denser Pareto front and a wider distribution compared to MOGA with ~ 44 optimal solutions compared to ~ 23 for MOGA. A competing behavior between the SPEA2 and NSGA-II is also observed. However, SPEA2 excelled in terms of the number of the optimal solutions ~ 50 compared to ~ 44 for the NSGA-II with a broader distribution. Table 1 summarizes the results of case 1.

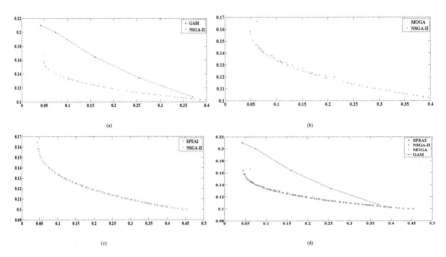

Fig. 2 The performance for case 1. a) NSGA-II vs. GAM, b) NSGA-II vs. MOGA, c) NSGA-II vs. SPEA2 and d) all algorithms.

Table 1 Number of the optimal solutions provided by each algorithm (case 1)

Algorithms	GAM	MOGA	SPEA2	NSGA-II
No. of optimal solutions	5*	23	50	44

* It requires executing the algorithm five times to acquire 5 optimal solutions.

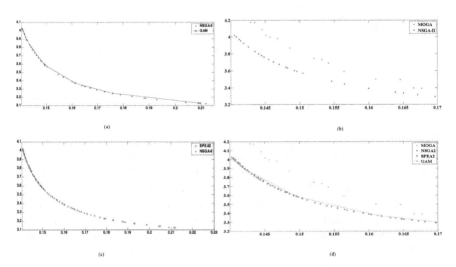

Fig. 3 The performance for case study 2. a) NSGA-II vs. GAM, b)NSGA-II vs. MOGA, c) NSGA-II vs. SPEA2 and d) all algorithms.

For case study 2, Figs 3(a), (b), (c) and (d) illustrate the comparison between NSGA-II and the GAM, MOGA and SPEA2 respectively.

GAM has a better performance (compared to case study 1) in terms of convergence with the defects of poor diversity and number of optimal solutions (~ 30 for NSGA-II compared to 5 for GAM) still present. The author attributes the acceptable convergence of GAM to the nature of the objective functions in case study 2. They are mathematically simpler than those of case study 1 (quadratic equations). NSGA-II outperformed MOGA in terms of the criteria used in this study to differentiate between the multi-objective algorithms i.e., the number of the optimal solutions (~30 for NSGA-II compared to ~15 for MOGA). An acceptable diversity is demonstrated by the MOGA compared to that in case study 1. MOGA is still poor in convergence towards the optimal Pareto front. This can be attributed to the fitness assignment method employed by MOGA which can cause slow convergence and occasionally poor diversity between optimal solutions.

A precise investigation of Fig. 3(c) showed that SPEA2 provided more optimal solutions compared to NSGA-II (~ 50 for SPEA2 compared to ~ 30 for NSGA-II). The diversity and convergence metrics showed an identical performance. This noticeable behavior of SPEA2 compared to NSGA-II is compatible with what have been seen in case study 1 when SPEA2 outperformed NSGA-II in terms of the number of the optimal solutions.

Fig. 3(d) represents a general performance comparison of the four algorithms for case study 2 optimization. In this case, the NSGA-II, SPEA2 and GAM exhibited adequate agreement while, MOGA performance did not. Results in Table 2 shows that SPEA2 can provide more optimal solutions compared to the others.

All in all, NSGA-II and SPEA2 outperformed their competitors in both case studies. If a choice is desired in the MCHS field, the balance will tend toward SPEA2 and NSGA-II with some preference for SPEA2.

Table 2 Number of the optimal solutions provided by each algorithm (case 2)

Algorithms	No. of optimal solutions
GAM	5
MOGA	15
SPEA2	50
NSGA-II	30

5   Conclusion

The overall performance of a MCHS has been optimized using four multi-objective optimization algorithms (GAM, MOGA, SPEA2 and NSGA-II). Generally, it can be concluded that SPEA2 and NSGA-II have outperformed the other algorithms in terms of number of the optimal solutions, faster convergence toward the Pareto optimal front and maintaining the required diversity among the optimal solutions. The distinctive result was that the SPEA2 has exhibited better performance than NSGA-II in terms of the number of optimal solutions. This finding opens the door for more investigations on multi-objective algorithms and provides more options on multi-objective algorithms that MCHS designers can utilize.

Acknowledgments. The authors would like to thank Universiti Teknologi Malaysia (UTM) under Grant University Project grant vote no. Q.K130000.2540.0H55 for the financial support provided throughout the course of this research.

References

1. Ahmed, M.A., Normah, M.G., Robiah, A.: Thermal and Hydrodynamic Analysis of Microchannel Heat Sinks: A Review. Renewable and Sustainable Energy Reviews (21), 614–622 (2013), doi:10.1016/j.rser.2013.01
2. Husain, A., Kim, K.Y.: Optimization of a microchannel heat sink with temperature dependent fluid properties. Applied Thermal Engineering (28), 1101–1107 (2008), doi:10.1016/j.applthermaleng.2007.12.001
3. Husain, A., Kim, K.Y.: Enhanced multi-objective optimization of a microchannel heat sink through evolutionary algorithm coupled with multiple surrogate models. Applied Thermal Engineering (30), 1683–1691 (2010), doi:10.1016/j.appithermaleng.2010.03.027
4. Gembicki, F.W.: Vector optimization for control with performance and parameter sensitivity indices. Ph.D. Thesis, Case Western Reserve Univ., Cleveland, Ohio (1974)
5. Zitzler, E., Laumanns, M., Thiele, L.: SPEA2: Improving the Strength Pareto Evolutionary Algorithm. Swiss Federal Institute of Technology (ETH), Zurich, Switzerland. Technical report TIK-Report103 (2001)
6. Ahmed, M.A., Normah, M.G., Robiah, A.: Optimization of an ammonia-coole3d rectangular microchannel heat sink using multi-objective non-dominated sorting genetic algorithm (NSGA2). Heat and Mass Transfer (2012), doi:10.1007/s00231-012-1016-8
7. Deng, B., Qi, Y.C., Kim, N.: An improved porous medium model for microchannel heat sinks. Applied Thermal Engineering (30), 2512–2517 (2010), doi:10.1016/j.applthermaleng.2010.06.025

Solution of the Inverse Stefan Problem by Applying the Procedure Based on the Modified Harmony Search Algorithm

Edyta Hetmaniok, Damian Słota, Adam Zielonka, and Roman Wituła

Abstract. In the paper we present an algorithm for solving the two-phase one-dimensional inverse Stefan problem with the temperature measurements given in selected points of the solid phase as an additional information. Proposed procedure bases on the modified Harmony Search algorithm and solving of the considered problem consists in reconstruction of the function describing the heat transfer coefficient.

1 Introduction

In the last decades a number of optimization algorithms imitating behaviors from the real world have found a great popularity. This group includes the Harmony Search algorithm proposed by Zong Woo Geem [1]. Idea of the Harmony Search algorithm is based on the similarity between the process of jazz improvisation and the problem of optimizing the function. Jazz improvisation consists in finding the best state of harmony, similarly as the optimization algorithm consists in finding the argument realizing minimum of the function. Referring to the problem of function optimization we can consider the arguments of function as the notes and the values for these arguments as the tones of instruments caused by these notes. And similarly like the musicians are searching for the combination of notes giving the best harmony of music, we are seeking the argument in which minimum of the function is taken. Classical version of HS algorithm has been already used by the authors for solving the inverse heat conduction problem in paper [4]. Received results appeared to be satisfying, however they gave an impulse to modify the algorithm in case when several successive executions of the algorithm do not improve the result.

In this paper we present a procedure based on the Harmony Search algorithm enabling to solve the two-phase axisymmetric one-dimensional inverse Stefan problem

Edyta Hetmaniok · Damian Słota · Adam Zielonka · Roman Wituła
Institute of Mathematics, Silesian University of Technology, Kaszubska 23, Gliwice, Poland
e-mail: {Edyta.Hetmaniok,Damian.Slota,Adam.Zielonka,
Roman.Witula}@polsl.pl

M. Ali et al. (Eds.): *Contemporary Challenges & Solutions in Applied AI*, SCI 489, pp. 175–180.
DOI: 10.1007/978-3-319-00651-2_24   © Springer International Publishing Switzerland 2013

describing the thermal process with the phase transition, solving of which in analytical way is impossible because of the missing boundary condition [3, 5, 6, 7, 8, 9]. The unknown input information is compensated by the additional information given by the temperature measurements taken in selected points of the solid phase. Solving of the considered problem consists in reconstructing the heat transfer coefficient appearing in boundary conditions, so that the temperature in the given points of the solid phase would have the closest values as possible to the known control values. Proposition of using the artificial intelligence algorithms imitating the natural behavior for solving that kind of problem is already presented by the authors in [2].

2   Idea of the Modified Harmony Search Algorithm

In details, the modified Harmony Search algorithm runs of the following way.

1. Initial data: minimized function $f(x_1, \ldots, x_n)$, range of the variables $a_i \leq x_i \leq b_i$, $i = 1, \ldots, n$, size of the harmony memory vector HMS $(1 - 100)$, harmony memory considering rate coefficient $HMCR$ $(0.7 - 0.99)$, pitch adjusting rate coefficient PAR $(0.1 - 0.5)$, number of iterations IT.
2. Preparation of the harmony memory vector HM – we randomly select HMS number of vectors $(x_1, \ldots, x_n)$ and we order them in vector HM according to the increasing values $f(x_1, \ldots, x_n)$:

$$HM = \begin{bmatrix} x_1^1, \ldots, x_n^1 & f(\mathbf{x}^1) \\ \vdots & \vdots \\ x_1^{HMS}, \ldots, x_n^{HMS} & f(\mathbf{x}^{HMS}) \end{bmatrix}.$$

3. Selection of the new harmony $\mathbf{x}' = (x_1', \ldots, x_n')$.
   For each $i = 1, \ldots, n$ the element x_i' is selected:
   - with probability equal to $HMCR$ from among numbers x_i collected in the harmony memory vector HM;
   - with probability equal to $1 - HMCR$ randomly from the range $a_i \leq x_i \leq b_i$.
     If in the previous step the element x_i' is selected from the harmony memory vector HM then:
   - with probability equal to PAR we modify the element x_i' in the following way: $x_i' \rightarrow x_i' + \alpha$ (we regulate the sound of note) for $\alpha = bw \cdot u$, where bw denotes the bandwidth – part of range of the variables and u is the randomly selected number from interval $[-1, 1]$;
   - with probability equal to $1 - PAR$ we do nothing.
4. If $f(\mathbf{x}') < f(\mathbf{x}^{HMS})$ then we put the element $\mathbf{x}'$ into harmony memory vector HM in place of the element $\mathbf{x}^{HMS}$ and we order vector HM according to the increasing values of minimized function.
5. If the successive 5 iterations do not bring any improvement of the result we upgrade the bandwidth (this step is a modification in relation to the classical HS algorithm [1] proposed by the authors):

$$bw \rightarrow 0.5 \cdot bw.$$

6. Steps 2–5 are repeated IT number of times. The first element of vector HM defines the solution.

3   Formulation of the Problem

Let us consider region $\Omega = [0,b] \times [0,t^*]$ divided into two subregions: Ω_1 taken by the liquid phase and Ω_2 taken by the solid phase. The interface (freezing front) Γ_g, is located along function $r = \xi(t)$. Boundary of region Ω is divided into five parts: $\Gamma_0 = \{(r,0) : r \in [0,b]\}$, $\Gamma_{11} = \{(0,t) : t \in [0,t_k)\}$, $\Gamma_{12} = \{(0,t) : t \in [t_k,t^*]\}$, $\Gamma_{21} = \{(b,t) : t \in [0,t_p)\}$ and $\Gamma_{22} = \{(b,t) : t \in [t_p,t^*]\}$. In selected points of the solid phase $((r_i,t_j) \in \Omega_2)$ the values of temperature are known:

$$T_2(r_i,t_j) = U_{ij}, \quad i = 1,\ldots,N_1, \quad j = 1,\ldots,N_2,$$

where N_1 denotes the number of sensors (thermocouples) and N_2 means the number of measurements taken from each sensor.

The problem consists in determining the function α defined on boundaries Γ_{2k} (for $k = 1,2$) such that the function ξ describing the interface position and the distributions of temperature T_k in regions Ω_k ($k = 1,2$), calculated for reconstructed α, would satisfy the axisymmetric one-dimensional heat conduction equation inside the regions Ω_k (for $k = 1,2$):

$$c_k \rho_k \frac{\partial T_k}{\partial t}(r,t) = \frac{1}{r}\frac{\partial}{\partial r}\left(\lambda_k r \frac{\partial T_k}{\partial r}(r,t)\right) \tag{1}$$

as well as the following initial and boundary conditions:

$$T_1(r,0) = T_0, \quad (T_0 > T^*) \quad \text{on } \Gamma_0, \tag{2}$$

$$\frac{\partial T_k}{\partial r}(0,t) = 0 \quad \text{on } \Gamma_{1k}, \ k = 1,2, \tag{3}$$

$$-\lambda_k \frac{\partial T_k}{\partial r}(b,t) = \alpha(t)\left(T_k(b,t) - T_\infty\right) \quad \text{on } \Gamma_{2k}, \ k = 1,2, \tag{4}$$

$$T_1(\xi(t),t) = T_2(\xi(t),t) = T^* \quad \text{on } \Gamma_g, \tag{5}$$

$$L\rho_2 \frac{d\xi}{dt} = -\lambda_1 \frac{\partial T_1(r,t)}{\partial r}\bigg|_{r=\xi(t)} + \lambda_2 \frac{\partial T_2(r,t)}{\partial r}\bigg|_{r=\xi(t)} \quad \text{on } \Gamma_g, \tag{6}$$

where c_k, ρ_k and λ_k are, respectively, the specific heat, mass density and thermal conductivity in liquid phase ($k = 1$) and solid phase ($k = 2$), α denotes the heat transfer coefficient, T_0 – the initial temperature, T_∞ – the ambient temperature, T^* – the solidification temperature, L describes the latent heat of fusion and, finally, t and r refer to the time and spatial location.

Direct Stefan problem described by equations (1)–(6) for the fixed form of heat transfer coefficient can be solved by using the alternating phase truncation method

[8, 9]. In result of this the course of temperature in solid phase can be determined. Values of temperature U_{ij}, calculated for the known exact values of sought coefficient, are considered as the exact values of temperature. By using the calculated temperatures T_{ij} and given temperatures U_{ij} the following functional is constructed:

$$J(\alpha) = \sum_{i=1}^{N_1} \sum_{j=1}^{N_2} \left(T_{ij} - U_{ij}\right)^2, \tag{7}$$

representing the error of approximate solution. By minimizing functional (7) with the aid of modified Harmony Search algorithm the values of parameter α assuring the best approximation of temperature can be found.

4 Numerical Verification

Elaborated procedure is tested by executing the experiment modeled with the aid of problem described by equations (1)–(6) for the following values of parameters:

$b = 0.08$ [m], $\lambda_1 = 104$ [W/(m K)], $\lambda_2 = 240$ [W/(m K)], $c_1 = 1290$ [J/(kg K)], $c_2 = 1000$ [J/(kg K)], $\rho_1 = 2380$ [kg/m^3], $\rho_2 = 2679$ [kg/m^3], $L = 390000$ [J/kg], solidification temperature $T^* = 930$ [K], ambient temperature $T_\infty = 298$ [K], initial temperature $T_0 = 1013$ [K]. Known exact form of the sought heat transfer coefficient α [W/(m^2 K)] for the considered process is given below:

$$\alpha(t) = \begin{cases} 1200 & \text{for } t \in [0,90), \\ 800 & \text{for } t \in [90,250), \\ 250 & \text{for } t \in [250,1000]. \end{cases}$$

For constructing functional (7) we use the exact values of temperature and values noised by the random error of 2 and 5%. The thermocouple is located in point $r = 0.07$ [m]. Direct Stefan problem associated with the investigated inverse problem is solved by using the finite difference method with applying the alternating phase truncation method for the mesh with steps equal to $\Delta t = 0.1$ and $\Delta r = b/500$. Modified Harmony Search algorithm is executed for size of the harmony memory vector $HMS = 15$, harmony memory considering rate $HMCR = 0.85$, pitch adjusting rate $PAR = 0.3$ and number of iterations $IT = 300$. Elements of the initial harmony memory vector HM are randomly selected from the range $[100, 2000]$ and the initial value of bandwidth parameter bw corresponds with 10% of the range of variables. Moreover, because of the heuristic nature of HS algorithm in each considered case the calculations were evaluated for 20 times and as the approximate values of reconstructed coefficient we accepted the best of obtained results.

Figure 1 present the relative errors of identification of the respective coefficients α_i, $i = 1, 2, 3$ in dependence on the number of iterations IT obtained for input data burdened by 5% error. One can observe that in presented execution of the procedure about 50 iterations is enough to obtain good results. The results stabilize on some satisfying level and further iterations do not give any significant improvement.

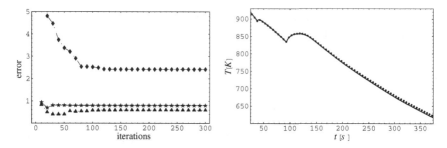

Fig. 1 Results for 5% input data error: left figure – relative errors of coefficients α_i reconstruction for the successive iterations ($\blacklozenge$ – for α_1, $\blacktriangle$ – for α_2, $\bigstar$ – for α_3), right figure – comparison of the exact (solid line) and reconstructed (dotted line) distribution of temperature in point $r = 0.07[m]$

Table 1 Mean values of reconstructed α_i ($i = 1, 2, 3$), standard deviations s_{α_i}, absolute and relative errors of this reconstruction, absolute and relative errors of temperature T reconstruction obtained in various iterations and noises of input data

noise	IT	i	$\overline{\alpha_i}$	s_{α_i}	$\delta_{\alpha_i}[\%]$	$\Delta_T[K]$	$\delta_T[\%]$
		1	1198.70	16.29	0.108		
	50	2	803.20	11.37	0.400	0.878	0.103
		3	252.80	5.97	1.119		
		1	1198.70	5.81	0.108		
2%	100	2	803.20	3.83	0.400	0.906	0.107
		3	252.94	5.08	1.176		
		1	1198.70	5.48	0.108		
	300	2	803.27	3.84	0.359	0.829	0.098
		3	252.80	4.45	1.118		
		1	1209.72	44.36	0.810		
	50	2	796.76	28.05	0.405	1.321	0.156
		3	241.57	7.50	3.371		
		1	1209.61	34.75	0.801		
5%	100	2	795.38	18.96	0.578	1.119	0.132
		3	243.70	6.08	2.519		
		1	1209.59	28.06	0.799		
	300	2	795.27	13.39	0.592	1.087	0.128
		3	243.96	5.97	2.416		

Stability of the procedure is confirmed as well by the results received for unburdened input data, errors of which converge very quickly to zero. Figure displays also the reconstructions of temperature distribution in measurement point $r = 0.07[m]$ compared with the known exact distribution. One can see that the reconstructed and known courses of temperature almost cover. Statistical elaboration of results obtained in 20 executions of the procedure for various numbers of iterations and for various noises of input data are compiled in Table 1. In each case the reconstruction errors are much smaller than input data errors and small values of standard deviations confirm stability of the procedure.

5 Conclusions

In this paper the method of solving the inverse Stefan problem by applying the modified Harmony Search algorithm as a tool of minimizing the appropriate functional is proposed. Presented results of the heat transfer coefficient reconstruction are good in each case of burdened input data. Satisfying results were obtained quite quickly, for small number of iterations and relatively small number of elements in the harmony memory vector, however improvement of the precision of the received results is impossible from a certain moment. Summarizing, the modified Harmony Search algorithm can be found as a useful tool for solving optimization problems of considered kind, more fast and efficient in comparison with classical methods.

Acknowledgements. This project has been financed from the funds of the National Science Centre granted on the basis of decision DEC-2011/03/B/ST8/06004.

References

1. Geem, Z.W., Kim, J.H., Loganathan, G.V.: A new heuristic optimization algorithm: Harmony Search. Simulation 76, 60–68 (2001)
2. Grzymkowski, R., Hetmaniok, E., Słota, D., Zielonka, A.: Application of the Ant Colony Optimization algorithm in solving the inverse Stefan problem. Steel Res. Int. special edition: Metal Forming, 1287–1290 (2012)
3. Gupta, S.C.: The Classical Stefan Problem. Basic Concepts. Modelling and Analysis. Elsevier, Amsterdam (2003)
4. Hetmaniok, E., Jama, D., Słota, D., Zielonka, A.: Application of the Harmony Search algorithm in solving the inverse heat conduction. Scientific Notes of Silesian University of Technology (Zesz. Nauk. Pol. Śl.), Applied Mathematics 1, 99–108 (2011)
5. Johansson, B., Lesnic, D., Reeve, T.: A method of fundamental solutions for the one-dimensional inverse Stefan problem. Appl. Math. Modelling 35, 4367–4378 (2011)
6. Liu, C.-S.: Solving two typical inverse Stefan problems by using the Lie-group shooting method. Int. J. Heat Mass Transfer 54, 1941–1949 (2011)
7. Ren, H.-S.: Application of the heat-balance integral to an inverse Stefan problem. Int. J. Therm. Sci. 46, 118–127 (2007)
8. Słota, D.: Solving the inverse Stefan design problem using genetic algorithm. Inverse Probl. Sci. Eng. 16, 829–846 (2008)
9. Słota, D.: Restoring boundary conditions in the solidification of pure metals. Comput. & Structures 89, 48–54 (2011)

Part IX
Robotics

Cascade Safe Formation Control for a Fleet of Underactuated Surface Vessels Using the DCOP Approach

Alejandro Rozenfeld, Jawhar Ghommam, Rodrigo Picos, and Gerardo Acosta

Abstract. This paper considers the formation control of multiple underactuated surface vessel. A distributed cooperative control using the relative information among neighboring vehicles is proposed such that the flock of multiple vehicles forms a desired geometric formation pattern whose center moves along a desired trajectory. In order to guarantee safe flock navigation and interaction of vehicles with the environment, we propose to extend the designed formation tracking controller to more sophisticated algorithm that prevent the vehicles from colliding with environmental obstacles with unknown sizes and locations based on a Decentralized Constrained Optimizing Problem "DCOP" strategy.

1 Introduction

Broad strategies for cooperative multi-agent missions have been studied extensively in the literature. The main existing approaches to vehicle formation are the behavioral approach [1], virtual structure technique [3], leader follower approach [2] and potential functions [4]. Each of the aforementioned approaches has its advantages and drawbacks. However one major common issue arises from the different

Alejandro Rozenfeld · Gerardo Acosta
INTELYMEC Group, CIFICEN-CONICET & Faculty of Engineering, National University
of Buenos Aires Province Center, Argentina
e-mail: alejandro.rozenfeld@gmail.com, ggacosta@fio.unicen.edu.ar

Jawhar Ghommam
The National School of Engineering of Sfax, Research Unit on Mechatronics and
Autonomous Systems, BP W, 3038 Sfax, Tunisia
e-mail: jawhar.ghommam@gmail.com

Rodrigo Picos
Universitat de les Illes Balears, Spain
e-mail: rodrigo.picos@uib.es

M. Ali et al. (Eds.): *Contemporary Challenges & Solutions in Applied AI*, SCI 489, pp. 183–188.
DOI: 10.1007/978-3-319-00651-2_25   © Springer International Publishing Switzerland 2013

approaches is formation stability, specifically when the system of vehicles evolves in disturbed environment which may accidently cause collisions among vehicles or the environment.

In this paper, we investigate new linear formation control for the navigation of multi-underactuated marine vehicle agents. Roughly speaking, there are two types of formation control objectives. The first control objective is to design cooperative controller such that the group of marine vehicles converge to a fixed geometric pattern with a desired orientation. The objective which is the key point for our formation strategy is to design distributed cooperative controllers such that the vehicles converge to some locations within the group formation to form a desired geometric pattern by ensuring all vehicles move along a predefined trajectory or path. In the recent literature, [2] proposed a general approach for leader following formation strategy. However this approach is only limited to stationary formation shape and doesn't allow changes in formation. Although this approach allows certain robustness of the formation to perturbations on the underactuated vehicles, its implementation is rather complex.

The contribution of this paper are twofold. The first contribution is to expand the H_∞ optimal control, to cooperative tracking controller with the aid of graph theory and notions from linear system theory. The idea behind our design lays in the use of an observer-type error consensus protocol to ensure fast convergence of the formation to the desired pattern. When the vehicles encounter obstructing objects on their trajectories, the formation shape is re-scaled to squeeze through the obstacles then restores back its original shape when no obstacles are detected. The scalability factors are parameters determined based on a decision maker that relies on an decentralized optimizing technique called DCOP [5].

The reminder of the paper is organized as follows: In the next section, the mathematical modeling of the USV is presented. Section 3, the cooperative tracking control problem formulation is presented. In Section 4, the main results for the Virtual Structure "VS" formation control is presented. Section 5, proposes a solution based on DCOP strategy for safe formation manoeuver. In section 6, numerical simulations are given. Finally in Section 7, some concluding remarks are provided.

2 Model of the USV

The group of marine vehicles that we are considering here are identical with three degrees of freedom each of which kinematic and dynamic model are given by the following equations [6]

$$\dot{x}_i = u_i \cos(\psi_i) - v_i \sin(\psi_i)$$

$$\dot{y}_i = u_i \sin(\psi_i) + v_i \cos(\psi_i)$$

$$\dot{\psi}_i = r_i$$

$$\dot{u}_i = \frac{m_{22i}}{m_{11i}} v_i r_i - \frac{d_{11i}}{m_{11i}} u_i + \frac{1}{m_{11i}} \tau_{ui} + w_{1i} \qquad (1)$$

$$\dot{v}_i = -\frac{m_{11i}}{m_{22i}} u_i r_i - \frac{d_{22i}}{m_{22i}} v + w_{2i}$$

$$\dot{r}_i = \frac{m_{11i} - m_{22i}}{m_{33i}} u_i v_i - \frac{d_{33i}}{m_{33i}} r_i + \frac{1}{m_{33i}} \tau_{ri} + w_{3i}$$

where x_i, y_i and ψ_i are the surge displacement, sway displacement and yaw angle in the earth fixed frame, u_i, v_i and r_i denote surge, sway and yaw velocities. The positive constant terms m_{lli}, $1 \leq l \leq 3$ denote the ship inertia including added mass. The positive constant terms d_{ll} represent the hydrodynamic damping in surge, sway and yaw. Finally, the available controls are the surge force τ_{ui} and the yaw moment τ_{ri}. The terms w_{1i}, w_{2i} and w_{3i} are disturbances that influence the acceleration of the vehicle.

3 Cooperative Control Problem Formulation

A group of N underactuated marine vessels are steered in such a way they move each to a given location in the formation. The vertices of the formation pattern are defined with respect to a virtual vehicle which model is identical to (1) which moves along a predefined trajectory $(x_d(t), y_d(t))$ and desired orientation ψ_d generated by simulating the vessel model described above without the disturbances. as given by the following equations. The coordinates $(p_{xi}, p_{yi}), i = 1, \ldots, N$ of the vertexes are defined with respect to a local system of coordinates associated to the virtual robot. The individual desired trajectory for each vertex is given by the following expression.

$$\begin{bmatrix} x_{di}(t) \\ y_{di}(t) \end{bmatrix} = \begin{bmatrix} x_d(t) \\ y_d(t) \end{bmatrix} + \begin{bmatrix} \cos\phi_i & -\sin\phi_i \\ \sin\phi_i & \cos\phi_i \end{bmatrix} \begin{bmatrix} p_{xi} \\ p_{yi} \end{bmatrix}$$

$$p_i^d = p_d + R(\phi_i) p_{xyi} \qquad (2)$$

where $p_i^d = [x_{di}, y_{di}]^\top, p_d = [x_d, y_d]^\top, p_{xyi} = [p_{xi}, p_{yi}]^\top$, $R(\bullet)$ is the rotation matrix and ϕ_i is a desired orientation angle specifying the orientation of a node robot with respect to the VS center. A scaled formation pattern is given with respect to a newly defined individual desired trajectory for vehicle i as

$$p_i^d = p_d + \frac{1}{\lambda_i} R(\phi_i + \theta_i) p_{xyi}$$

$$\theta_{di} = \theta_d \qquad (3)$$

where $\lambda_i \in \mathbb{R}^+$ is a constant scalar parameter and θ_i is a parameter angle that varies in $[0, \pi]$, both parameters would allow the scalability of the formation pattern. Our aim is to design the appropriate control inputs τ_{ui} and τ_{ri} to satisfy the following limits:

$$\lim_{t \to \infty} (x_{ei} - x_{ej}) = 0, \lim_{t \to \infty} (y_{ei} - y_{ej}) = 0, \lim_{t \to \infty} (\psi_{ei} - \psi_{ej}) = 0, \quad \forall i \neq j \qquad (4)$$

4 Cooperative Control Design

The control design is divided into two parts. The first one consists in designing an H^∞ optimal controller by minimizing a given performance index to ensure the tracking issue of the control objective while attenuating the effect of disturbances on the systems. The second part of the overall cooperative controller incorporates interaction terms in the local controller "part one" to ensure that each member of the team has knowledge of the status of other members.

4.1 Local Controller

In this subsection, in order to simplify the subsequent development a compact representation for error tracking dynamics of the underactuated marine vessel is developed based on (1). The error tracking equation of motion for the vehicle is given by

$$\dot{\mathbf{p}}_{ei} = \mathbf{F}_{ei}(t, \mathbf{p}_{ei}) + B_i(t)\mathbf{u}_{ei} + D_i(t)\mathbf{w}_i(t) \qquad (5)$$

where $\mathbf{u}_{ei} \in \mathbb{R}^2$ is the control input vector to be designed to ensure tracking and synchronization performances. Toward this purpose, the control vector input is split into two components as follows:

$$\mathbf{u}_i = \mathbf{u}_{1i} + \mathbf{u}_{2i} \qquad (6)$$

As mentioned before the first component serves to minimize the subsequent performance index for the tracking problem issue and is given as follows

$$\mathbf{u}_{1i} = \mathbf{u}_{di} - B_i^\top X_i \mathbf{p}_{ei} \qquad (7)$$

The next step of the design is to include some terms to define the dependency of the control input of agent i on it's neighbors' information, this will be the issue of the next subsection.

4.2 Global Control Design

In this step we suggest to include a cooperative term to the initial design to account for the interaction among the vehicle using graph theory [7] as follows

$$\zeta_i = \sum_{j \neq i} a_{ij} C_i (\mathbf{p}_{ei} - \mathbf{p}_{ej}) \tag{8}$$

where a_{ij} is an element of the adjacency matrix associated to the graph $\mathcal{G}$ and $C_i = [1,1,1,0,0,0]^\top$. The cooperative control that we propose that completes the H^∞ optimal control design, takes the following form

$$\mathbf{u}_{2i}(\chi_j) = K_i v_i$$
$$\dot{v}_i = \left(\mathbf{A}_{ei} - B_i B_i^\top X_i + B_i K_i \right) v_i + \alpha \mathcal{D} \left(\sum_{j \neq i} a_{ij} C_i (v_i - v_j) - \zeta_i \right) \tag{9}$$

where $v_i \in \mathbb{R}^6$ is the integral protocol state, $K_i \in \mathbb{R}^{6 \times 6}$ is a positive gain matrix. The positive definite matrix $\mathcal{D} \in \mathbb{R}^{6 \times 6}$ is the feedback cooperative gain that gives insight on the coupling strength among the vehicles and α is a positive constant.

5 Formation Shape Transformation Using the DCOP Strategy

In this section the DCOP approach is used as a decision maker for the formation pattern reconfiguration, it provides directives to the underactuated vessels to change and reach their desired location within the formation. Assume that vehicle "1" is the first one detecting the presence of an obstacle in the trajectory, it triggers a collective task aiming to construct a Priority Spanning Tree (PST). This PST will be used to modify the formation pattern into a line in order to cross the tunnel. Once the PST achieves a steady state according to the distances between vehicles, the formation is ready to reorganize its pattern.

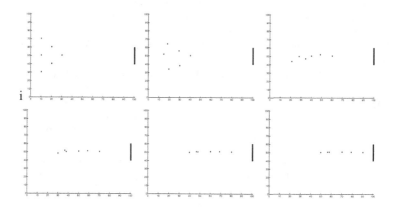

Fig. 1 a sequence of snapshots from a simulation applying the DCOP strategy

6   Numerical Simulations

In this section, simulation results are presented to validate the effectiveness of the proposed controller in combination with an intelligent optimizing algorithm "The DCOP". We consider a group of six underactuated marine vehicles. In the simulations, the underactuated vehicles are required to move from an initial configuration "diamond formation pattern", to enter a narrow passage with an aligned configuration by following a straight line. Figure 1 shows a sequence of snapshots of the vehicles' motion trying to reconfigure their locations in the formation geometric pattern.

7   Conclusion

In this paper a H_∞ optimal cooperative controller has been developed for the virtual structure formation strategy for a team of underactuated marine vehicles. The idea behind the cooperative control design lays in the use of an observer-type error consensus protocol to ensure fast convergence of the formation to the desired pattern. Intelligent algorithm based on the DCOP strategy has been proposed to enable the vehicles to change their location within the formation team when they encounter obstructing obstacles like entering a tunnel. Simulations have been carried out on a group of 6 underactuated marine vehicles moving in diamond formation then changing into an inline formation as though they are entering a narrow passage.

References

1. Balch, T., Arkin, R.C.: Behavior-based formation control for multirobot teams. IEEE Transactions on Robotics and Automation 14(6), 926–939 (1998)
2. Cuia, R., Ge, S.S., How, B.V.E., Choob, Y.S.: Leader-follower formation control of underactuated autonomous underwater vehicles. Ocean Engineering 37(17-18), 1491–1502 (2010)
3. Ghommam, J., Mehrjerdi, H., Saad, M., Mnif, F.: Formation path following control of unicycle-type mobile robots. Robotics and Autonomous Systems 58(5), 727–736 (2010)
4. Do, K.D.: Output-feedback formation tracking control of unicycle-type mobile robots with limited sensing ranges. Robotics and Autonomous Systems 57(1), 34–47 (2009)
5. Maheswaran, R.T., Pearce, J.P., Tambe, M.: Distributed algorithms for DCOP: a graphical-game-based approach. In: PDCS (2004)
6. Fossen, T.I.: Guidance and control of ocean vehicles. John Wiley & Sons (1994)
7. Godsil, C., Royle, G.: Algebraic Graph Theory. Graduated Texts in Mathematics (2001)

UMH's Navigation in Unknown Environment Based on Pre-planning Guided Fuzzy Reactive Controller

Xuzhi Chen, Zhijun Meng, Wei He, and Kaipeng Wang

Abstract. Based on the sparse A* search (SAS) algorithm and the fuzzy reactive controller (FRC), we propose a novel method of navigation for unmanned helicopter (UMH). SAS is applied to plan a path based on the understanding of pre-known obstacles and threats. Then, UMH travels along the path. The FRC, which employs Mamdani fuzzy methodology and pre-planning guidance, monitors the flight process and react in real-time to keep flight safety. Simulations show that this approach can find out the global optimal path and realize dynamic navigation for UMH.

1 Introduction

To realize autonomy for UMH, path planning technology was developed. Taking terrain, obstacle and target in consideration, path planner generates a global optimal trajectory [1]. But when the task is carried out in an unknown environment, the trajectory may need to be revised continuously. In this case, traditional path planning method can be prohibitively expensive to compute. Reactive method is another way to realize UMH's autonomy. It uses simple formulas to tackle the real-time challenge, but cannot guarantee an appropriate solution to every possible situation due to the uncertainty of the external environment [2]. In this paper, combining the advantages of the above methods, a novel navigation method based on path planning and reactive obstacle avoidance is proposed.

Geometrically, path planning methods can be categorized as graph methods and grid methods. Graph algorithms, such as Voronoi diagram methods [3] and probabilistic roadmap methods (PRM) [4], are based on skeleton diagrams. While, grid methods like SAS [5] et al. are based on cell decomposition of the configuration space (C-space). In this paper, we use SAS as the path pre-planning method.

Xuzhi Chen · Zhijun Meng · Wei He · Kaipeng Wang
School of Aeronautic Science and Engineering, Beihang University, Beijing, China

M. Ali et al. (Eds.): *Contemporary Challenges & Solutions in Applied AI*, SCI 489, pp. 189–194.
DOI: 10.1007/978-3-319-00651-2_26   © Springer International Publishing Switzerland 2013

After pre-planning, UMH flies along the obtained trajectory. To avoid the newly encountered obstacle or threat, reactive method like CD* algorithm [6] and vector field histogram (VFH) [7] method must be applied. In our approach, we use Mamdani fuzzy method to build a reactive obstacle avoidance controller. We then carried out a series of improvements to implement it in UMH's navigation.

2 Basic Assumptions

Several assumptions were made in this article.

- First, UMH is assumed to fly at a constant altitude. In this manner, the configuration space (C-space) is a two dimension one.
- Second, obstacles are treated as no-fly zones, denoted as NFZs which must be avoided. Thus, their cost values (CV) are set to be positive infinity.
- Third, threats are characterized by their radar detector range. Thus, their CVs are computed by the simplified radar function:

$$CV = \frac{k_T}{R^4} \tag{1}$$

where, k_T is the coefficient of the threat and R is the distance to the threat.

- Fourth, it is assumed that the situation awareness system is fully aware of any threat, no matter whether it is static or dynamic.
- Fifth, when pre-planning, it is only the static threats that are taken into account. Dynamic threats will be detected and handled in obstacle avoidance process.

3 Pre-planning Based on SAS

The SAS is a variation of the standard A* search, which is used quite extensively in route planning applications. Compared to standard A* search, SAS saves tremendous time and computing space as it takes into account various mission constraints to prune the search space. In our research, there are three constraints to be applied: Minimum route leg length constraint, Maximum turn angle constraint and Route distance constraint.

The pre-planning in our research follows closely the exposition in *Robust Algorithm for Real-Time Route Planning* [5]. Due to the limitation of the article's length, the SAS's calculation process is not detailed here.

4 Design of FRC

4.1 FRC Based on Mamdani Method

Fuzzy logic allows for values to be represented by degrees of truth [8]. In this way, the knowledge and experiences of human operators can be included. As such, the fuzzy logic methodology lends itself well to the problem of goal-seeking and reactive obstacle avoidance. The whole process contains three stages: fuzzification of inputs, fuzzy reasoning and defuzzification of fuzzy command.

UMH use onboard sensors to monitor the surrounding environment and detect newly emerging obstacle. The sensors output the distance $d_{o,i}$ and deviating angle $\psi_{o,i}$ of each obstacle which are fed to FRC. The distance d_G and deviating angle ψ_G of goal are also fed to FRC.

Two separate FRCs are combined to accomplish goal-seeking and obstacle avoidance. The goal-seeking controller and the obstacle avoidance controller cooperate with each other to navigate UMH.

The inputs used for each of these control schemes are $d_{o,i}$, $\psi_{o,i}$, d_G and ψ_G. For each distance, the linguistic values are small (S), medium (M) and large (L) and the medium crisp values are 10 meters, 20 meters and 40meters. For each angle, the linguistic values are negative big (NB), negative small (NS), zero (Z), positive small (PS) and positive big (PB) and the medium crisp values are $-20°$, $-10°$, $0°$, $10°$ and $20°$. Gauss function is chosen as the membership function.

Although UMH can fly very slowly or even hover, it is still better that UMH maintains a constant speed. So, the outputs of both FRCs are yaw rate. For yaw rate, the linguistic values are negative big (NB), negative small (NS), zero (Z), positive small (PS) and positive big (PB) and the medium crisp values are $-10°/s$, $-5°/s$, $0°/s$, $5°/s$ and $10°/s$. Gauss function is chosen as the membership function.

Inference matrices for the goal-seeking and obstacle avoidance behaviors are defined in Table 1 and Table 2, respectively.

Table 1 Goal-seeking fuzzy rules

		Deviating Angle of Goal				
		NB	NS	Z	PS	PB
Distance	S	PB	PB	Z	NB	NB
	M	PB	PS	Z	NS	NB
	L	PS	PS	Z	NS	NS

Table 2 obstacle avoidance fuzzy rules

		Deviating Angle of Threat				
		NB	NS	Z	PS	PB
Distance	S	NB	NB	PB	PB	PB
	M	NS	NB	PB	PB	PS
	L	NS	NS	PS	PS	PS

According to Mamdani theory [9], centriod method is applied here to realize defuzzification, which is computed as follow:

$$u^* = \frac{\int u(x_i)x_i dx_i}{\int u(x_i)dx_i}$$

(2)

where, u^* is the crisp output, $u(x_i)$ is the membership degree at x_i.

4.2 Dynamic Rolling Window Limitation

In a flight, UMH may encounter many obstacles. If all detected obstacles are fed to FRC, it will greatly increase computational burden of FRC and introduces unnecessary maneuvers. So, FRC's input area is limited according to the following rules: Obstacle which is relatively far from UMH or deviates greatly from the UMH's current heading is unlikely to collide with UMH in a short term and thus ignored for now. Therefore, a fan-shaped FRC's input area, named dynamic rolling window (DRW), is applied to filter the detected obstacles for FRC. As illuminated in Fig. 1, UMH flies along a pre-planned path and obstacles no. 1~5 are all new and not considered in pre-planning. It is only obstacle no. 2 that is inputted to FRC because it is the only one inside of DRW.

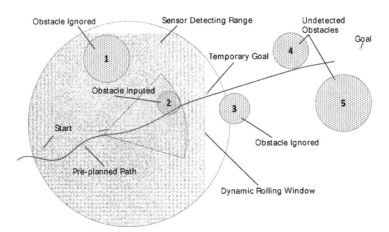

Fig. 1 A sample of dynamic rolling window

4.3 Pre-planning Guided

Since the pre-planned path contains mission information and global understanding which are not considered by FRC, it is suggested that UMH flies close to the pre-planned path. Therefore, a temporary goal is fed to FRC, which is intersection of the DRW's edge and the pre-planned path, as shown in Fig. 2.

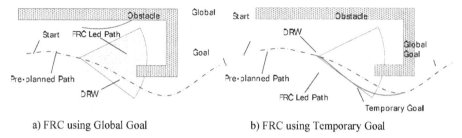

a) FRC using Global Goal   b) FRC using Temporary Goal

Fig. 2 A sample shows the benefit of pre-planning guided FRC

One example of the benefit of this approach is shown in Fig. 2. In (a) the FRC uses global goal in its guidance. Since the DRW is small, FRC only considers a part of the whole "C" shaped obstacle. Together with the contribution of global goal, FRC may lead UMH into a cul-de-sac. In (b), the FRC selects its temporary goal along the pre-planned path and then successfully bypass the obstacle.

Note that the pre-planning guidance rule has to be given up when newly appearing obstacle changes the environment so greatly that UMH has to fly to new region very far from the pre-planned path.

5   Simulation

A series of simulations were carried out in Matlab to verify the proposed method. Two typical results are presented in Fig. 3, where red area indicates obstacles and yellow area indicates threats. UMH starts from the left bottom aiming at the goal which lies at the right top. DP in the figures means "detecting point" where UMH detects the pre-unknown obstacle or threat.

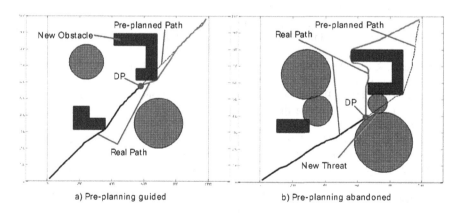

a) Pre-planning guided   b) Pre-planning abandoned

Fig. 3 Navigation in unknown environment

In (a), a new "C" shaped obstacle which is not included in pre-planning appears in the right top part. Under the guidance of pre-planned path, FRC navigates UMH safely and efficiently bypass the obstacle and reach the goal. In (b), UMH detects the unforeseen threat (the little yellow circle in the middle of the right part) at DP, and then initials the avoidance maneuver. As the pre-planned path is completely blocked, FRC does not follow it this time.

6   Conclusion

A comprehensive structure of pre-planning guided FRC was designed, implemented, and tuned to safely navigate UMH in unknown environment to reach a desired goal without collision. The shortages of only using path planning or reactive avoidance along are discussed. Then, we proposed the novel approach to take advantages of the above two method. Simulation results show that our approach is fully capable for the navigation in unknown environment for UMH.

By assuming UMH travels at a constant height, we simplified the problem to a 2D space. But in actual application, this simplification is impractical and may cause serious problems. So, in future works we should extend the method to 3D.

References

1. Tisdale, J., Kim, Z.W., Hedrick, J.K.: Autonomous UAV path planning and estimation: An online path planning framework for cooperative search and localization. IEEE Robotics and Automation Magazine 16(2), 35–42 (2009)
2. Scherer, S., Singh, S., Chamberlain, L., Elgersma, M.: Flying Fast and Low Among Obstacles: Methodology and Experiments. Int. J. Robot. Res. 27(5), 549–574 (2008)
3. Liu, Z., Shi, J.G., Gao, X.G.: Application of voronoi diagram in flight path planning. Acta Aeronautica et Astronautica Sinica 29(suppl.), 15–19 (2008) (in Chinese)
4. Park, J.J., Kim, J.H., Song, J.B.: Path planning for a robot manipulator based on probabilistic roadmap and reinforcement learning. Int. J. Control Autom. Syst. 5(6), 674–680 (2007)
5. Szczerba, R.J., Galkowski, P., Glickstein, I.S., Ternullo, N.: Robust Algorithm for Real-Time Route Planning. IEEE Trans. Aero. Elec. Sys. 36(3), 869–878 (2000)
6. Stenza, A.: CD*: a Real-time resolution optimal re-planning for globally constraint problem. In: Proceedings of the 18th National Conference Artificial Intelligence, pp. 1088–1096 (2002)
7. Borenstein, J., Koren, Y.: Thevector field histogram–Fast obstacle avoidance for mobile robots. IEEE Trans. Rob. Autom. 7(3), 278–288 (1991)
8. Yen, J., Langari, R.: Fuzzy Logic: Intelligence, Control, and Information. Prentice Hall (1999)
9. Mandani, E.K.: Application of Fuzzy Logic to Approximate Reasoning, Using Linguistic Synthesis. Queen Mary College (1974)

Part X
Special Session on Decision Support for Safety-Related Systems

Developing Context-Free Grammars for Equation Discovery: An Application in Earthquake Engineering

Štefan Markič and Vlado Stankovski*

Abstract. In the machine-learning area of equation discovery (ED) context-free grammars (CFG) can be used to generate equation structures that best describe the dependencies in a given data set. Our goal is to investigate the possible strategies of incorporating domain knowledge into a CFG, and evaluate the effect on the obtained results in the ED process. As a case study, the Lagramge ED system is used to discover equations that predict the peak ground acceleration (PGA) in an earthquake event. Existing equations for PGA represent rich domain knowledge and are used to form three different CFGs. The obtained results demonstrate that the inclusion of domain knowledge in the CFG which is neither too general, neither too specific, may lead to new, high-precision equation models for PGA.

Keywords: equation discovery, Lagramge, context-free grammar, domain knowledge, earthquake engineering, peak ground acceleration.

1 Introduction

Equation discovery (ED) is a sub area of machine learning aiming at automatic induction of mathematical models expressed as equations. The goal is to find an equation structure from a given set of operators, functions and variables that represents an appropriate model for the provided data set. ED systems like Lagramge[1] use the context-free grammar (CFG) formalism to restrict the hypothesis space of

Štefan Markič · Vlado Stankovski
Faculty of Civil and Geodetic Engineering, University of Ljubljana, Slovenia, Jamova cesta 2, SI-1000 Ljubljana
e-mail: `vlado.stankovski@fgg.uni-lj.si`

* Corresponding author.

[1] The Lagramge release 2.2 used in this study is available as open-source software at URL: `http://www-ai.ijs.si/~ljupco/ed/lagrange.html` (accessed 6th February 2012)

M. Ali et al. (Eds.): *Contemporary Challenges & Solutions in Applied AI*, SCI 489, pp. 197–203.
DOI: 10.1007/978-3-319-00651-2_27   © Springer International Publishing Switzerland 2013

$$\log_{10}(PGA) = 1.04159 + 0.91333 \times M_w - 0.08140 \times M_w^2$$
$$+ (-2.92728 + 0.28120 \times M_w) \times \log_{10} \sqrt{R_{jb}^2 + 7.86638^2}$$
$$+ \begin{cases} 0.08753 \text{ if} & V_{s,30} < 360 \\ 0.01527 \text{ if } 360 \le V_{s,30} < 800 \\ 0 & \text{if } 800 \le V_{s,30} \end{cases} + \begin{cases} -0.04189 \text{ if F} = 0 \\ 0 & \text{if F} = 0.5 \\ 0.08015 \text{ if F} = 1 \end{cases} \quad (1)$$

$$\ln(PGA) = f(M_w, R_{jb}, V_{s,30}, F) \quad (2)$$

possible equation structures [4, 6, 8]. Usually, this is achieved by incorporating domain knowledge in the productions of the CFG. The construction of a CFG, however, requires considerable know-how, as also noted by [4], and may range from more general to more explicit specification of existing equation structures. The goals of the present study are therefore: *i*) to investigate the possible ways of forming the CFGs; *ii*) to compare the various ways of inclusion of domain knowledge; and *iii*) to observe the effects on the obtained results, by following a motivating example.

Case Study. In civil engineering an important task is to properly design a structure, bearing in mind that a devastating earthquake could occur during its lifetime. The ground-motion prediction equations help the structural engineer to estimate the possible earthquake load by providing the correlation between seismically important variables, (e.g., peak ground acceleration (PGA)) and significant seismological aspects (e.g., magnitude and distance) [2]. An example of a modern equation from [1] is presented in Eq. (1), and the ED problem is formulated as Eq. (2).

Data Set. The PF-L data set used in this study consists of 3550 earthquake recordings and is taken from the study of [7]. The data set is very sparse at high magnitudes and short distances. The independent variables used in this study are similarly to [6, 1, 7]: *i*) the moment magnitude M_w; *ii*) the source-to-site Joyner-Boore distance R_{jb} (*km*); *iii*) the average soil shear-wave velocity in the upper 30 meters of soil $V_{s,30}$ (*m/s*); and *iv*) the style-of-faulting F with values of *a*) $F = 0$ for normal; *b*) $F = 0.5$ for strike-slip; and *c*) $F = 1$ for reverse faults. The dependent variable is PGA (*g*-units), defined as the geometrical average of both horizontal components.

2   The Lagramge System

The Lagramge ED system [8] takes as input two input files: a data set and a CFG. The data set consists of a table of measurements of dependent $ln(PGA)$ and independent variables $M_w, R_{jb}, V_{s,30}, F$. The $CFG = \{N, T, P, S\}$ prescribes the syntax of an equation. First, it contains finite disjunctive sets of non-terminals N and terminals T. The terminals are all the independent variables and a special symbol *const*, which is explained under the parameter fitting paragraph. The most important part of the CFG are the productions $P = \{P_1, P_2, \ldots, P_n\}$, which denote the grammatical rules that relate the non-terminals among themselves and to the terminals. The standard form of a production P is $A \rightarrow \alpha$, where $A \in N$, $\alpha \in N \cup T$. The operators or

$$MSE = \frac{1}{n} \sum_{i=1}^{n} (PGA - \widehat{PGA})^2 \qquad (3)$$

functions used can be already or user-defined in the programming language C. The Lagramge system uses the annotation with the logical *or* operator $A \rightarrow \alpha_1 | \alpha_2 | \dots | \alpha_n$ for productions $A \rightarrow \alpha_1, A \rightarrow \alpha_2, \dots, A \rightarrow \alpha_n$. Finally, $S \in N$ is a special non-terminal symbol, from which the derivation of the expressions starts.

Parameter Fitting. During the derivation process, Lagramge continuously applies productions to all the non-terminals until all the symbols in the expression (formula structure) are terminals. Such an expression contains one or more special terminal symbols *const*, the syntax of which is [*name* : *lowest value* : *starting value* : *highest value*]. A non-linear fitting method, either Downhill Simplex or Levenberg-Marquardt, is used to determine the values of these symbols. The fitting procedure can be repeated (by setting up the Lagramge parameter m). In the ED process the Lagramge system minimizes the value of the Mean-Squared Error (MSE) function given in Eq. (3), in which n is the number of records and PGA and $\widehat{PGA}$ are the measured and predicted values of the PGA, respectively.

Search Strategies. The Lagramge system provides two search strategies: *i*) an exhaustive search strategy, where all possible equation structures in the hypothesis space are fitted; and *ii*) a heuristic, also called beam search strategy, according to which one can set the number of equations saved in each production step with the value of the input parameter beam width b.

3 Approaches to the CFG Definition

Three unique CFGs were defined, each with a different level of incorporation of domain knowledge that also takes the form of equations systematised by [3]. The defined CFGs are presented in the following paragraphs along with a description of the rationale of the approaches. As the ratio between fault types F is not known and the variable $V_{s,30}$ is often divided into classes, two conditional functions *ifl* and *ife* were defined that compare two values for their smallness or equality, respectively. To improve the readability of the grammars twelve auxiliary productions were defined: *i*) Ma, Ra, Vs and Fa productions lead to the variables' addresses known to the Lagramge algorithm, e.g. Ma $\rightarrow$ variable_M; *ii*) K0, K1, K2, K180, K360, K750, K800 productions lead to presumed constant values, e.g. K0 $\rightarrow$ const[_:0:0:0]; and *iii*) the production Ko $\rightarrow$ const[_:-100:0.1:100] denotes the symbol *const* fitted to the data which is limited to values between -100 and 100 based on the literature review [3]. The definitions and productions were included in all CFGs and are presented in Table 1A.

General CFG. With the CFG provided in Table 1B very diverse equation structures can be built and tested, hence, it is named General. It contains all the different functions and operators used in already existing equations, which can be combined together in all possible ways with the possibility of recursion. The exponent in the

Table 1 *ife* and *ifl* functions, auxiliary productions and designed CFGs

A) *ifl* and *ife* functions and auxiliary productions	double ifl(double val, double comp, double t, double f) { Ma → variable_M
	     return((val < comp) ? t : f); } *...similar...*
	double ife(double val, double comp, double t, double f) { K0 → const[_:0:0:0]
	     return((val == comp) ? t : f); } *...similar...*

B) General CFG	A → A + A | (A) × (A) | (A) / (A) | pow(A, const[_:0:0.1:5]) | exp(A) | log(A) | Ma | Ra | Vs | Fa | Ko

C) Specialized CFG – Eq. (1) as illustrative example	E → Ko + Ko × Ma + Ko × pow(Ma, K2)
	     + (Ko + Ko × Ma) × log(sqrt(pow(Ra, K2) + Ko))
	     + ifl(Vs, K360, Ko, ifl(Vs, K750, Ko, K0))
	     + ife(Fa, K0, Ko, ife(Fa, K1, Ko, K0))

D) Intermediate CFG	Eq → Ko + FM + FR + FV + FF
	FM → (FM + Ko × FM1) | Ko × FM1
	FM1 → Ma | pow(Ma, K2) | pow(Ma + Ko, const[_:1:1.5:5]) | exp(Ko × Ma)
	FR → Ko × FM1 × FR1 | FR + Ko × FR1 | Ko × FR1
	FR1 → ln(Ra + Ko) | ln(Ra + Ko × FM1) | ln(pow(Ra, K2) + Ko) |
	     ln(pow(Ra, K2) + Ko × FM1) | pow(Ra + Ko, -K1) | pow(Ra + Ko, -K2)
	FV → FM1 × FV1 | FV1
	FV1 → K0 | ifl(Vs, K180, Ko, ifl(Vs, K360, Ko, ifl(Vs, K800, Ko, K0))) |
	     Ko × ln(Vs/const[_:0:800:4000]) | ifl(Vs, const[_:0:800:4000] , Ko, K0)
	FF → ife(Fa, K1, Ko, ife(Fa, K0, Ko, K0)) | K0

power term is limited between values of 0 and 5, as negative powers are not needed because the production A → (A)/(A) can generate them and powers greater than 5 were not seen in [3].

Specialized CFG. For the second CFG all the existing equation structures developed for the PGA modeling published by European authors from these subsections of Section 2 of the study [3] were transcribed in grammar productions: 12, 16, 18, 22, 23, 34, 35, 40, 46, 50, 59, 67, 72, 74, 76, 84, 86, 88, 92, 102, 108, 113, 118–120, 124, 128, 146, 152, 157, 165, 175, 179, 181, 187, 189, 191, 192, 195, 197, 198, 202, 205–211, 235, 239, 242, 254, 256, 260, 263, 266, 275, 276, 277, 282, 283, 288 and 289. An explicit use of the depth variable h was substituted with a *const* parameter, as the PF-L database does not include such information. The resulting Specialized CFG (see Table 1C) includes all-together 62 different equation structures from 64 published articles and is not included here due to lack of space.

Intermediate CFG. The design of the third CFG was one of the most difficult tasks undertaken. The actual productions were defined by abstracting the formulae copied for the Specialized CFG and provide the combinatorial freedom of the General CFG. The use of this CFG (see Table 1D) first leads from the root symbol Eq to non-terminal functions FM, FR, FV and FF, named after the dependence they model, e.g., FM for $f(M_w)$. Each of these functions can then be succeeded with their own special sub-functions gathered during the literature review [3]. The Intermediate

$$\ln(PGA) = -0.563429 \times (8.8822 + 2 \times M_\text{w} + R_\text{jb}^{0.392507} + \ln(M_\text{w})) \tag{4}$$

$$\ln(PGA) = -4.78208 + 1.90683 \times M_\text{w} - 0.152475 \times M_\text{w}^2$$
$$+ (-2.20764 + 0.169162 \times M_\text{w}) \times \ln\sqrt{R_\text{jb}^2 + 61.1355} \tag{5}$$

$$+ \begin{cases} 0.490597 & \text{if} & V_{\text{s},30} < 180 \\ 0.297185 & \text{if } 180 \le V_{\text{s},30} < 360 \\ 0.0577997 & \text{if } 360 \le V_{\text{s},30} < 750 \\ 0 & \text{if } 750 \le V_{\text{s},30} \end{cases} + \begin{cases} 0.07482 & \text{if } F = 0 \\ 0 & \text{if } F = 0.5 \\ 0.09791 & \text{if } F = 1 \end{cases}$$

$$\ln(PGA) = 1.23491 - 0.117808 \times (M_\text{w} - 6.60126)^2$$
$$- 0.18992 \times \exp(-0.12945 \times M_\text{w} + 1.8984) \times \ln(R_\text{jb}^2 + 57.2879) \tag{6}$$
$$- 0.310872 \times \ln\frac{V_{\text{s},30}}{464.231} + \begin{cases} 0.0951288 & \text{if } F = 0 \\ 0 & \text{if } F = 0.5 \\ 0.0720131 & \text{if } F = 1 \end{cases}$$

$$\ln(PGA) = 4.5735 - 1.6929 \times M_\text{w} + 0.2417 \times M_\text{w}^2 - 6.6761 \times \exp(-7.6020 \times M_\text{w})$$
$$- 0.0091837 \times \frac{\exp(1.3707 \times M_\text{w})}{R_\text{jb} + 100} - 1.6782 \times \ln(R_\text{jb} + 12.7587) \tag{7}$$
$$- 0.291666 \times \ln\frac{V_{\text{s},30}}{4000} + \begin{cases} 0.1254 & \text{if } F = 0 \\ 0 & \text{if } F = 0.5 \\ 0.1188 & \text{if } F = 1 \end{cases}$$

CFG limits the space of possible equations to only those that are the most plausible according to the studied domain knowledge.

4 Results

Based on some initial trial and error experiments using the recently developed Web application [5] it was decided to explore the hypothesis space: *i)* of the General CFG with beam search $d = 7, b = 50$; *ii)* of the Specialized CFG with exhaustive search; and *iii)* of the Intermediate CFG with exhaustive search $d = 4$ and beam search $d = 10, b = 50$. The parameter m for fitting restarts was set to 50. The PF-L data set was preprocessed by converting the PGA into their logarithmic values and randomly split 10 times in a 90 % to 10 % proportion, with the purpose of a 10-fold cross validation. The best equations found from all four experiments are: *i)* Eq. (4) for the General CFG; *ii)* Eq. (5) for the Specialized CFG; *iii)* Eq. (6) for the Intermediate CFG with exhaustive search; and *iv)* Eq. (7) for the Intermediate CFG with beam search. The averages and standard deviations of the MSE criterion on all testing data sets for the Eqs. (1), (4), (5), (6) and (7) are shown in Table 2.

Table 2 Calculated averages and standard deviations of the MSE criterion on the testing data sets for Eq. (1), (4), (5), (6) and (7)

Equation	(1)	(4)	(5)	(6)	(7)
$\overline{MSE}$	0.4595	0.4569	0.4138	0.4135	0.3957
σ_{MSE}	0.0257	0.0296	0.0244	0.0245	0.0236

5 Conclusions

In this study, the Lagramge ED system was used to induce equations that predict the earthquake's PGA. In the past decades many authors have addressed this problem and developed numerous equations, which represent rich and specific domain knowledge. The existing domain knowledge was formalised in the CFGs to find new, potentially more accurate models by following three different approaches. The first approach (General CFG) took into account only the basic functions and operators which are present in the existing equations. The second approach (Specialized CFG) took into account the exact formulae taken from 64 articles of European authors [3]. The third approach (Intermediate CFG) was a combination of the previous two and took into account only the most often modeled variables dependencies in existing equations, but allowing them to be combined freely.

Our investigation shows that the use of domain knowledge contributes to the discovery of more precise equation models for PGA. The inclusion of strict equation structures in Specialized CFG provided a 10% reduction of the MSE criterion when compared to General CFG. However, the combination of both approaches in Intermediate CFG performed even better providing a 15% reduction. We conclude that careful definition of grammar productions defines an infinite, but quality hypothesis space and may lead to obtaining the best results.

Acknowledgements. The authors are grateful to Iztok Peruš and Peter Fajfar for providing the PF-L data set and fruitful discussions. Special thanks go to Ljupčo Todorovski for guidance when using the Lagramge ED system.

References

1. Akkar, S., Bommer, J.J.: Empirical equations for the prediction of PGA, PGV, and spectral accelerations in Europe, the Mediterranean region, and the Middle east. Seismol. Res. Lett. 81(2), 195–206 (2010)
2. Douglas, J.: Earthquake ground motion estimation using strong-motion records: a review of equations for the estimation of peak ground acceleration and response spectral ordinates. Earth-Sci. Rev. 61(1-2), 43–104 (2003)
3. Douglas, J.: Ground-motion prediction equations 1964–2010. Final report, BRGM/RP-59356-FR and PEER/2011/102, Pac. Earthq. Eng. Res. Cent., 444 p. (2011)
4. Kompare, B., Todorovski, L., Džeroski, S.: Modelling and prediction of phytoplankton growth with equation discovery: case study–Lake Glumsø, Denmark. Verh. Int. Verein. Limnol. 27, 3626–3631 (2001)

5. Markič, Š., Dirnbek, J., Stankovski, V.: A Grid Application for Equation Discovery in the Earthquake Engineering Domain. In: Proc. of the 3rd Int. Conf. on Parall, Distrib., Grid and Cloud. Comp. for Eng. Civil-Comp Press (in press, 2013)
6. Markič, Š., Stankovski, V.: An equation-discovery approach to earthquake-ground-motion prediction. Eng. Appl. Artific. Intel. (in press, 2013), doi:10.1016/j.engappai.2012.12.005
7. Peruš, I., Fajfar, P.: Ground-motion prediction by a non-parametric approach. Earthq. Eng. & Struct. Dyn. 39, 1395–1416 (2010)
8. Todorovski, L., Džeroski, S.: Declarative bias in equation discovery. In: Proc. of the 14th Int. Conf. on Mach. Learn., pp. 376–384 (1997)

Neural Networks to Select Ultrasonic Data in Non Destructive Testing

Thouraya Merazi Meksen, Malika Boudraa, and Bachir Boudraa

Abstract. In recent years, research concerning the automatic interpretation of data from non destructive testing (NDT) is being focused with an aim of assessing embedded flaws, quickly and accurately in a cost effective fashion. This is because data yielded by NDT techniques or procedures are usually in the form of signals or images which often do not present direct information of the structure's condition. Signal processing has provided powerful techniques to extract the desired information on material characterization and defect detection from ultrasonic signals. The imagery available can add additional and significant dimension in NDT information. The task of this work is to minimize the volume of data to process replacing ultrasonic images type TOFD by sparse matrix, as there is no reason to store and operate on a huge number of zeros, especially when large structures are inspected. A combination of two types of neural networks, a perceptron and a Self Organizing Map (SOM) of Kohonen is used to distinguish between a noise signal from a defect signal in one hand, and to select the sparse matrix elements which correspond to the locations of the defects in the other hand. This new approach to data storage will provide an advantage for the implementations on embedded systems as it allows the normalization of the sparse matrix by fixing its dimension.

1 Introduction

The use of non destructive testing allows the analysis of structures internal properties without causing damage to the material. Various methods have been developed to detect defects in structures and to evaluate their locations, sizes and characteristics. Some of these methods are based on analysis of the transmission of

Thouraya Merazi Meksen · Malika Boudraa · Bachir Boudraa
University of Science & Technology Houari Boumediene
BP 32, El Alia, Bab Ezzouar, 16111, Algiers, Algeria
e-mail: tmeksen@usthb.dz

M. Ali et al. (Eds.): *Contemporary Challenges & Solutions in Applied AI*, SCI 489, pp. 205–210.
DOI: 10.1007/978-3-319-00651-2_28   © Springer International Publishing Switzerland 2013

different signals such as ultrasonics, acoustic emission, thermography, x-radiography, eddy current. In the last decade, ultrasonic techniques are becoming an effective alternative to radio-graphic tests. X-ray widely used to detect and sizing discontinuities, presents the disadvantage to produce ionizing radiation and needs to develop a film, which takes some times to provide results [1].

When large structures are inspected, operators have to acquire and interpret large volumes of complex data. So, automated signal analysis systems are finding increasing applications in a variety of industries where the diagnostics is difficult. Ultrasonic data can be displayed as images and can add additional and significant dimension in NDT information. Many advanced image processing algorithms have provided powerful techniques to extract from ultrasonic images the desired information on sizing and defect detection [2,3,4]. But all these methods require considerable amount of computation, making them difficult for real-time operations. Many mechanized inspection techniques, sensors, and systems for automating defect detection and location have been developed [5,6,7]. However, the location and sizing of a defect is an almost entirely manual process: The operator will mark on the scan, using a mouse, where the component echoes lie, and thus where defect lies. The apparatus will then perform the correction and give an indication of the defect size according to what has been indicated by the operator. Hardwares have been developed and tools of image processing are implemented in integrated circuits in order to completely automate the control [8,9].

The aim of this work is to minimize the data to store and to process in order to save memory and computational time. An original approach for data acquisition and representation, which consists on sparse matrix construction instead of an ultrasonic image type TOFD (Time Of Flight Diffraction) is described. It is based on the TOFD technique but avoids the image formation. The sparse matrix is built by combination of a perceptron and a Self Organizing Map algorithm (SOM) of Kohonen in order to select a defined number of samples from the signals. The perceptron will discriminate during the scanning, between signals reached by a defect and signal constituted only by noise. The SOM normalizes the amount of data to a fixed number, after the storage of the pertinent samples co-ordinates in the sparse matrix format. Section 2 of this paper describes ultrasonic non destructive inspection and TOFD technique. In section 3, the two types of Neural Networks used in this work are developed, namely perceptron and Self Organizing Map of Kohonen. Measurements and results are described in section 4. Section 5 concerns the conclusion.

2   Time of Flight Diffraction Technique

The basic components of an ultrasonic inspection system are a pulser/receiver, cabling, transducers, and acoustic wave propagation and scattering. The pulser section of the pulser/receiver generates short electrical pulses which travel through the cabling to the transmitting transducer. The transducer converts these electrical pulses into acoustic pulse at its acoustic output port, which can be or not be in contact with the material under control. This ultrasonic beam is also transmitted into the solid component being inspected and interacts with any flaw that is present. The flaw generates scattered wave pulses travelling in many directions, and some of these pulses reach the receiving transducer which converts them into electrical pulses. These electrical pulses travel again through cabling to the receiver section of the pulser/receiver, where they are amplified and displayed.

TOFD technique uses the travel time of a diffracted wave at the tip of a discontinuity [10]. Two transducers, one as a transmitter and the second as a receiver are moved automatically step by step according to a straight line and the diffracted signals are recorded and displayed as images. Those images provide different texture patterns for the detected defects and automatic texture segmentation is investigated using different techniques to improve the detection and classification of defects.

In his thesis, Sallard demonstrates that a generator of a hole, can be assimilated to a top of a crack [11]. So, a test block containing a hole has been used in this work to test this method (figure 1). Figure 2 shows the result obtained scanning a test block containing an artificial crack. Every row is constituted of a samples of a reached signal. White pixels correspond to positive amplitudes and grey ones correspond to zero level. The Zone of Interest is limited to the arc, formed by successive signals received during the linear displacements of the probes.

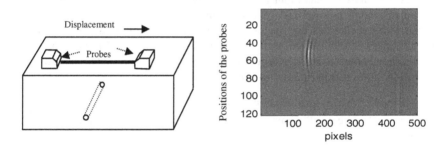

Fig. 1 Test block with an artificial defect   **Fig. 2** TOFD image showing a defect

This image is formed by signals containing 500 samples, at 120 positions of the probes. It will contain 120x500 pixels, while the zone of interest is relatively very small.

3   Artificial Neural Networks

In an Artificial Neural Network structure, many simple nonlinear processing elements, called neurons, are interconnected via weighted synapses to form a network inputs [12]. Learning process is of different types: supervised learning, unsupervised learning, self-organized learning. In a supervised approach, the network is fed with necessary input and the appropriate output for the specified inputs is given. The output is achieved together with a global error function. The computed output is compared to the desired output to evaluate the performance of the neural network. The computed error function is then used to update the weights with an aim of achieving output that is close to the desired one.

The Perceptron is a binary classifier that maps its input x (a real valued vector) to an output single binary value. If two sets are linearly separable, this classification can be used to decide whether a given vector belongs to one class or another. The function of each neuron is to compute a weighted sum of all synapse inputs, add the sum to a predefined bias and pass the result through a nonlinear sigmoidal (threshold) function whose output ranges between 0 and 1. In this work the learning stage of the perceptron had been made using two types of signals: Signals with defect signatures and signals constituted only by structural noise in order to discriminate those two types.

Unsupervised learning or self organizing learning does not require any assistance of desired outputs or an external teacher. Instead, during the training session, the neural network receives a number of different patterns and discovers significant features in these patterns and learns how to classify input data into appropriate categories.

Kohonen Self Organizing Maps is a model based on the idea of self organized or unsupervised learning [13]. The problem that data visualization attempts to solve is about reducing dimensions by producing maps of usually one or two dimensions which plot the similarities of the data by grouping similar data items together The network has input and output layers of neurons that are fully interconnected among themselves. At each step of training phase an output layer's neuron with weights that best match with input data (usually in a minimum Euclidian distance) is proclaimed as the winner. The weights of this neuron and its neighborhood neurons are then adjusted to be closer to the presented input data.

The algorithm is described as follows:

1- Initialize W with uniform-random values.

2- For each input vector P, compute the distance d between the vector P and the M weight vectors:

$$d = (P_k - W_l)^2 \qquad\qquad l = 1, 2, 3, \ldots M \qquad\qquad (1)$$

3- Select the vector Wx such that Wx satisfies Equation 2 :

$$(P_k - W_X)^2 = \min(P_k - W_l)^2 \qquad l = 1, 2, 3, \ldots M \qquad\qquad (2)$$

4- Update W_x using Equation 3:

$$W_x(t+1) = W_x(t) + \alpha(X_k - W_x) \qquad 0 < \alpha < 1 \qquad\qquad (3)$$

5- Go to step 2 until $W_l \approx P_l$ \qquad\qquad\qquad $l = 1, 2, 3 \ldots M.$

In the early learning stage, α is set about 0.8. As the learning progresses, α gradually becomes closer to 0.

4   Measurements and Results

According to the principle of the TOFD technique, two probes (2 Mhz) are moved step by step, by 5 mm each time, straight a line. At each position, the reached diffracted signal is first analyzed by a perceptron in order to determine if it is a "defect signal" or "a noise signal". From every signal of defect-signal class, the coordinates p_i and p_j corresponding respectively to the time of flight and the position of the probe, according to the straight line of the displacement, are stored. At the end of this process, a set of points $P(p_i, p_j)$ are determined and their number equals the number of the signals with defect signature.

In the next step, the Self Organization Map of Kohonen is applied with those points as inputs in order to reduce their number to a defined one depending on the desired sparse matrix dimension (30 in this work). The map will be a 1-dimensional layer of 30 neurons. For a learning rate $\alpha=0.7$, the convergence is obtained after 20 iterations.

Figure below shows the positions of the output elements resulting, obtained using signals that form the TOFD image on figure 2.

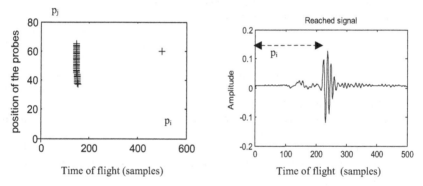

Fig. 3 Sparse Matrix obtained instead of the   **Fig. 4** Selected sample on the signal
TOFD image in fig.2

Instead of 120x500 pixels of the TOFD image, this method selects 30 elements which are sufficient to describe the pattern presented in the image. This number is defined by the Self organizing Map algorithm outputs and is independent of the quantity of initial data. The economy of memory is important and the defect location and characterization will be faster when analyzing only the sparse matrix elements.

5   Conclusions

In this work, a method to detect and locate cracks by analyzing a sparse matrix built from TOFD signals has been described. A first layer of a neural networks selects a point from the reached signal if a signal with defect signature   is presented in the zone of interest. The co-ordinates of this point which correspond respectively to the probe position and the time of flight of the signal are stored and used as inputs for a self organizing map network. Outputs will represent a group of points corresponding to the defect presented in the structure keeping the topology of the shape relative to the detected defect. Pattern recognition algorithms can be exploited, as it has. It is well known that crack defects produces parabolic forms on TOFD images. This work is the first step before implementation of such tools on programmable circuit for situations requiring near real-time processing.

References

1. Verkooijen, J.: TOFD to replace radiography. Insight 37(6), 433–435 (1995)
2. Chen, C.H.: Advanced Image Processing Methods for Ultrasonic NDE Research. In: World Cong. of Non Des. Testing, Proc. WCNDT 2004, Montreal, August 30-September 3, pp. 39–43 (2004)
3. Baskaran, G., Balasubramaniam, K.: Ultrasonic TOFD Flaw Sizing and Imaging in Thin Plates Using Embedded Signal Identification Technique (ESIT). Insight 2, 537–542 (2004)
4. Jasiuniene, E.: Ultrasonic Imaging Techniques for Non Destructive Testing of Nuclear Reactors, cooled by liquid Metals: Review. Ultragras 62(3), 39–43 (2007)
5. Cchatzakos, P., Markopoulos, Y.: Towards Robotic Non Destructive Inspection of Industrial Pipelines. In: 4th Int. Conf. on NDT, HSNDT 2007, Chania-Crete, Greece, October 11-14 (2007)
6. Martin, J., Gonzalez Bueno, R.: Ultrascope TOFD : Un sistema compacto para captura y procesamiento de imagenes TOFD. In: IV Conferencia Panamericana de END, PANNDT 2007, Buenos Aires, Aregentina, October 22-26 (2007)
7. Berke, M., Kleinert, W.D.: Portable Work Station for Ultrasonic Weld Inspection. In: 15th World Conf. of Non Destructive Testing, WCNDT 2000, Roma, Italy (2000)
8. Johnston, C.J., Gribbon, K.T.: Implementing Image Processing Algorithms on FPGAs. In: 11th Electronic New Zeland Conference, ENZCon 2004, Palmerston North, New Zeland, pp. 118–123 (2004)
9. Nelson, A.E.: Implementation of Image Processing Algorithm on FPGA Hardware. Thesis in Electrical Engineering, Faculty of the graduate school of Vanderbilt (2000)
10. Silk, M.G.: The Use of Diffraction-Based Time of Flight Measurements to Locate and Size Defects. British Journal of NDT 26, 208–213 (1984)
11. Sallard, J.: Etude d'une Méthode de Déconvolution Adaptée aux Images Ultrasonores. Thesis présented at the Institut National Polytechnique de Grenoble, France (1999)
12. Rosenblatt, F.: Principles of neurodynamics. Spartan, New York (1962)
13. Kohonen, T.: Self Organization and Associative Memory. Springer, Heidelberg (1988)

Part XI
Special Session on Innovations in Intelligent Computation and Applications

Stairway Detection Based on Extraction of Longest Increasing Subsequence of Horizontal Edges and Vanishing Point

Kaushik Deb, S.M. Towhidul Islam, Kazi Zakia Sultana, and Kang-Hyun Jo*

Abstract. Detection of stair region from a stair image is very crucial for autonomous climbing navigation and alarm system for blinds and visually impaired. In this regard, a framework is proposed in this paper for detecting stairways from stair images. For detection of the stair region, a natural property of stair is utilized that is steps of a stair appear sorted by their length from top to bottom of the stair. Based on this idea, initially, horizontal edge detection is performed on the stair image for detecting stair edges. In second step, longest horizontal edges are extracted from the edge image through edge linking. In third step, longest increasing subsequence (LIS) algorithm is applied on the horizontal edge image for extracting stair edge. Finally, the vanishing point is calculated from these sets of horizontal lines to confirm the detection of stair candidate region. Various stair images are used with a variety of conditions to test the proposed framework and results are presented to prove its effectiveness.

1 Introduction

In an unknown environment, navigation of an autonomous system is relatively easier when it moves on a plane surface such as roads, building floors, corridors etc. However, it becomes difficult when there is a stair in front of it. To navigate through the stair, it is first required to detect the stair area within the stair image. It is also important for alarm systems for blind and visually impaired. The proposed framework

Kaushik Deb · S.M. Towhidul Islam · Kazi Zakia Sultana
Dept. of CSE, Chittagong University of Engineering & Technology,
Chittagong-4349, Bangladesh
e-mail: debkaushik99@cuet.ac.bd

Kang-Hyun Jo
Dept. of EE and Information Systems, University of Ulsan, 680-749 Ulsan, South Korea
e-mail: acejo@ulsan.ac.kr

* Corresponding author.

M. Ali et al. (Eds.): *Contemporary Challenges & Solutions in Applied AI*, SCI 489, pp. 213–218.
DOI: 10.1007/978-3-319-00651-2_29   © Springer International Publishing Switzerland 2013

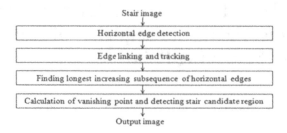

Fig. 1 The proposed stair detection framework

for stair detection is mainly based on a very important and unique property of a stair that is stair steps appear gradually increasing from top to bottom of the stair in a parallel arrangement. Because of the perspective nature of human vision, every stair shows this property when it is seen from a little distance from its front side. These steps appear as parallel lines in the edge image. Those parallel lines are required to be extracted for detection of the stair candidate region. This set of parallel lines is detected by using LIS algorithm. This is the key part of the proposed stair detection framework. From these parallel lines, the vanishing point is calculated to confirm the detection of stair candidate region.

2 The Proposed Stair Detection Framework

In this section, the four primary stages of the proposed stairway detection framework, i.e., horizontal edge detection, edge linking and tracking, finding longest increasing subsequence (LIS), and calculation of vanishing point and detecting stair candidate region have been discussed in details. In Fig. 1, the steps of our proposed framework are shown.

2.1 Horizontal Edge Detection

The input stair image is converted into gray scale image. The following steps are per-formed on this gray scale image to extract the horizontal edges.

Smoothing: In this step, noises and shadow effects are removed from the gray scale image for detection of the true edges.

Finding Edge Gradient Magnitude: The proposed method requires the horizontal edges mainly. Hence, Sobel operator is used in this step to estimate the gradient magnitude in the y-direction.

Non-maximum Suppression: After finding out the edge gradient magnitudes, non-maximum suppression is applied. Using the gradient magnitude value at every pixel of the image, it is determined whether a particular pixel is maxima or not. A threshold value is used in this step whose value is determined by

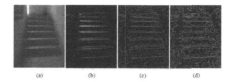

(a)   (b)   (c)   (d)

Fig. 2 Horizontal edge detection process: (a) input image, (b) gradient magnitude, (c) thin horizontal edges after non-maximum suppression, and (d) better horizontal edges after thresholding.

$1.2 * MEAN\_VALUE$ where $MEAN\_VALUE$ is the mean of all the gradient magnitude values.

Thresholding: In this step, we grow up the edges by performing another thresholding with $LOW\ THRESHOLD$ value. The value for $LOW\ THRESHOLD$ is determined by $0.6 * MEAN\_VALUE$. The horizontal edge detection process is pictorially shown in Fig. 2.

2.2 Edge Linking and Tracking

An image containing horizontal edges is mainly considered at this stage. As there are some noise and very small length horizontal edges in the image, edges with lengths greater than a certain threshold (threshold value = IMAGE WIDTH / 5) are extracted. In this step, the length and endpoints of each extracted edges are stored for further use.

2.3 Finding Longest Increasing Subsequence of Horizontal Edges

In this step, we utilize a very important and unique property of stair- stair steps appear gradually increasing from top to bottom of the stair in a parallel arrangement. Hence, these steps appear as an increasing sequence of parallel edges in the edge image. More precisely, if we have N horizontal edges in the current edge image, there are M gradually increasing edges where $M <= N$. This problem is similar to finding longest increasing subsequence from a sequence of numbers. Longest increasing subsequence can be solved using dynamic programming. This can be solved in $O(n^2)$ or $O(nlogn)$ time. Here, the only thing to consider is how we manipulate edges instead of numbers. In case of two numbers i and j, we assume $i > j$, if $|i| > |j|$ (where, (where, $|i|$ = magnitude of i and $|j|$ = magnitude of j). But in case of edges, edge i is greater than j if all the conditions - $Lx(i) < Lx(j)$, $Rx(i) > Rx(j)$, $Ly(i) > Ly(j)$ and $Ry(i) > Ry(j)$ are satisfied by edge i and edge j. Implementation of the LIS algorithm on a sample image is shown in Fig. 3.

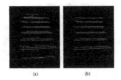

Fig. 3 Extracted LIS of edges a) Long horizontal edges (21 edges) and b) Potential stair edges extracted by LIS algorithm (11 edges)

2.4 Calculation of Vanishing Point and Detecting Candidate Region

The basic detection of staircase is finished already. This section is performed for the validation of detection. The vanishing point decides whether the detected parallel horizontal lines are extracted from a stair or from any other stair like object. Vanishing point can be defined as an imaginary point where several lines from an object converge. In case of stair, the point where the two handrails of the stair intersect can be defined as its vanishing point. Here, the left handrail is the straight line passing through left endpoints of the uppermost and lowermost edge in the LIS edge image. Similarly, right handrail is the straight line passing through right endpoints of the uppermost and lowermost edge in the LIS edge image. Hence, the vanishing point is the intersection point of these two straight lines.

After calculating the VP, we check its y co-ordinate. Vanishing point of a stair should not be located too high or too low. Most of the stairs vanishing point will reside inside the range $-2H < VP < 0$ where H is the height of the image and VP is the y co-ordinate of the vanishing point. If the calculated vanishing point is inside this range, the detected parallel lines certainly indicate a stair. Otherwise, these may be part of some other stair like object. As for example, rail lines have similar property as stairs. The position of a railway line is fully horizontal with respect to ground. Hence, vanishing point of a railway line is located inside the image. Fig. 4(b) shows experiment on a rail line image. The calculated vanishing point is (180, 40). The y co-ordinate of the vanishing point is 40. This positive y value distinguishes this image from a stair image. On the other hand, stairs have a slope with respect to ground. That's why stair's vanishing point resides outside the image. If the calculated vanishing point is outside the image i.e. inside the range $-2H < VP < 0$, we extract the stair region bounded by the left handrail, the uppermost edge (Edge 1 of Fig. 4(a)), the right handrail and the lowermost edge (Edge N of Fig. 4(a)).

3 Experimental Result

In this section, we show some experiments on some real stair images. All experiments were done on Intel 2160 @ 1.79 GHz processor with 2 GB RAM. All processing was done on 400x400 images. Processing was implemented by using C language with OpenCV. Fig. 5 shows various input images, intermediate processing

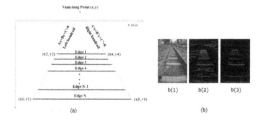

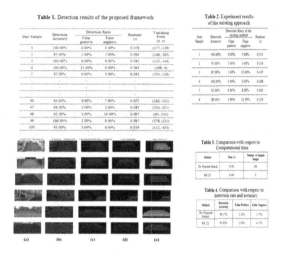

Fig. 4 Processing example: (a) Calculation of vanishing point (VP), and (b) Illustration of extracting long horizontal edges from rail line image: b1) a rail line image, b2) Extracting long horizontal edges from rail line image, and b3) LIS of edges

Fig. 5 Illustration of stair region segmentation: a) a stair image, b) edge image, c) horizontal edge image, d) LIS of horizontal edges, and e) extracted stair region

and correspond-ing output images. Table I shows the runtime, calculated vanishing point, accuracy of detection and detection rates of the stairs of the image database. The database which has been used during the experiments contains 100 stair images. Those images con-tain various indoor and outdoor stair images captured at various illumination conditions.

The runtime is due to performing horizontal edge detection, edge linking and tracking, finding LIS of edges and calculation of vanishing point. Each of these sample stairs has a vanishing point with negative y co-ordinate value. Stair sample 1 and 2 of Fig. 5 are straight forward stair images with good illumination conditions. Stair sample 3 and 6 shows results on indoor stairs at lower illumination conditions. Stair sample 4 and 5 shows experiment on clipped stairs with noisy background where some part of stair is cropped. All these varieties of staircases are extracted successfully by the proposed framework with acceptable accuracy and detection rates as shown in Table 1.

The proposed method has been compared with an existing method which has been presented in [2]. Table 2 shows the experimental results of this existing method of [2]. This table shows the detection accuracy, detection rates and the runtimes of the exist-ing approach. From these two tables we can compare the performance of the proposed method with the existing approach. The comparison based on the execution time or runtime has been shown in Table 3. This table shows average computational times. The proposed method's average computational time to process 100 images has been shown. Besides this, the existing method's average computational time to process 6 images has been shown also. The comparison with respect to average detection accuracy and rate has been shown in the Table 4.

4 Conclusion

In this paper a new algorithm for extracting a stair region from an image has been introduced. This work is important for autonomous systems to perform climbing process in unknown environments. The proposed algorithm was successfully tested on a group of stair images with varying styles and structures. These images were taken in various illumination conditions. The proposed algorithm successfully extracted stair regions from these images. In future, this work can be extended to get relevant information about a stair such as the distance of the stair from the camera, number of steps of the stair, height of each step of the stair etc. All these information are very important for climbing process of autonomous systems in unknown environments.

References

1. Hernández, D.C., Jo, K.H.: Stairway Segmentation Using Gabor Filter and Vanishing Point. In: Proc. IEEE Int. Conf. on Mechatronics and Automation, August 7-10 (2011)
2. Hernández, D.C., Jo, K.H.: Outdoor Stairway Segmentation Using Vertical Vanishing Point and Directional Filter. In: The 5th International Forum on Strategic Technology (2010)
3. Cong, Y., Li, X., Liu, L., Tang, Y.: A Stairway Detection Algorithm based o Vision for UGV Stair Climbing. In: IEEE International Conference on Networking, Sensing and Control (2008)
4. Se, S., Brady, M.: Vision-based detection of staircases. In: Fourth Asian Conference on Computer Vision, ACCV 2000, vol. 1, pp. 535–540 (2000)
5. Coremen, T.H.: Introduction to algorithms, 2nd edn., pp. 350–356
6. McLean, G.F., Kotturi, D.: Vanishing point detection by line clustering. IEEE Transactions on Pattern Analysis and Machine Intelligence 17(11), 1090–1095 (1995)
7. Canny, J.: A Computational Approach to Edge Detection. IEEE Trans. Pattern Analysis and Machine Intelligence 8(6), 679–698 (1986)

A Heuristic to the Multiple Container Loading Problem with Preference

Tian Tian, Andrew Lim, and Wenbin Zhu

Abstract. In this paper, we address the Multiple Container Loading Problem with Preference (MCLPP). It is derived from the real problems proposed by an audio equipment manufacturer. In the MCLPP, the numbers of various types of boxes can be adjusted based on box preferences. We need to add or delete boxes in a restricted way so that the ratio of the total preference of boxes to the total cost of containers is maximized. We develop a three-step search scheme to solve this problem. Test data is modified from existing benchmark test data for the multiple container loading cost minimization problem. Computational experiments show our approach is able to provide high quality solutions and they satisfy the need of the manufacturer.

1 Introduction

Our team cooperated with an international audio equipment manufacturer to solve problems emerged in the logistics process. One task is to design loading plans for packing various products into containers with the minimum cost. When packing products of an order into selected containers, the manufacturer found two common scenarios. In one scenario, all containers are fully filled except one whose utilization is small. In other words, a cheaper set of containers can substitute the original one such that all containers in this set are fully utilized with only a few products left unpacked. In the other scenario, only one container is not full. But its utilization is large, i.e., if several products are added it will be filled to its full capacity. In both scenarios, the ratio of the total preference of boxes to the total cost of containers

Tian Tian · Andrew Lim
Department of Management Sciences, City University of Hong Kong, Tat Chee Ave,
Kowloon Tong, Hong Kong SAR
e-mail: {tiantian,limandrew}@cityu.edu.hk

Wenbin Zhu
Department of Computer Science and Engineering, Hong Kong University of Science and
Technology, Clear Water Bay, Kowloon, Hong Kong SAR
e-mail: i@zhuwb.com

M. Ali et al. (Eds.): *Contemporary Challenges & Solutions in Applied AI*, SCI 489, pp. 219–224.
DOI: 10.1007/978-3-319-00651-2_30   © Springer International Publishing Switzerland 2013

(RPC) will probably be increased if the order can be slightly changed. We denote the
two scenarios as the MCLPP-Add (MCLPP-Add) and the MCLPP-Delete (MCLPP-
Delete) separately in the remaining part of this paper.

Fortunately, in the audio equipment manufacturer's case, slightly changing the
order is considered to be acceptable. There are several reasons. Firstly, customers
usually decide their orders based on forecasts for the future demand. However, no
forecasting is accurate. Secondly, market demand is always fluctuating which provides room for changes. Last but not least, customers who purchase from the manufacturer regularly may compensate the changes to the current order in future ones.
Therefore, it is meaningful to provide suggestions to customers on whether and how
to adjust the orders such that the unit shipping cost is minimized.

Usually, large deviations from the original order is inadmissible from both the
customers' and the manufacturer's point of view. The adjustment criteria can be described as follows. (1) Either a few products are added in or some items are deleted
from the original order. When adjusting an order, it is not allowed to add some types
of products and delete other types of products simultaneously. Furthermore, products to be added must belong to types that have been ordered by the customer. (2)
The total cost of containers for the new order should not be greater than that for the
original order. (3) Products to be added or deleted are selected according to their
preferences. In practice, the shipping of some items may be more desirable than
that of others. To measure the difference we introduce a preference indicator for
each type of products. The higher the preference is the more desired the product is.

We name the above problem as the Multiple Container Loading Problem (MCLP)
with Preference. In the MCLPP, we need to adjust the customer order in a restricted
way and generate loading plans for the new order. Our objective is to maximize
the ratio of the total preference of boxes to the total cost of containers. A simple
three-step heuristic is developed to solve the MCLPP according to the problem. To
test our algorithm, We modify the 350 benchmark MCLCMP instances proposed by
Che et al. [1]. Experimental results show that our heuristic method is able to provide
a reasonable solution to the MCLPP.

2 A Three-Step Heuristic Method

The solution technique proposed for the MCLPP in this study is a three-step heuristic approach (TS-heuristic) as given in Algorithm 2. We first try to find a cheap container combination to load all the boxes in the original customer order, *BoxList*. This
process is completed using a greedy method. Initially, we select a set of containers
and try to pack all boxes in *BoxList* into them. If all boxes are feasibly loaded, we
may find a smaller container combination that can handle *BoxList*. We keep looking
for smaller container combinations until no one can be found to load all boxes in
BoxList. If not all boxes in *BoxList* can be loaded into the initial container combination, we select a new container combination with larger volume. Similarly, the
process ends when we find a set of containers that can load all the boxes in *BoxList*.
At the end of the algorithm, a cheap and small container combination will be

reported and denoted as *cscc*. At the same time we record the packing patterns for the original order as *origSol*. Actually, the problem we need to solve in the first step is similar to the Multiple Container Loading Cost Minimization Problem (MCLCMP) except that in our problem the container combination should be the smallest in total volume. There are several papers talking about the MCLCMP and readers who are interested are referred to Eley et al. [2], Che et al. [1], Wei et al. [3] and Zhu et al. [4].

In the second step, we try to load more boxes into the container combination *cscc*. A binary search method (Algorithm 1) is called to determine the new set of boxes, *moreBoxList*.

Algorithm 1. Add Boxes into the Customer Order

ADDBOX(*BoxList, CC*)

```
 1   lVol = the total volume of containers in CC
 2   uVol = the total volume of boxes in BoxList
 3   sBoxVol = the smallest volume of boxes in BoxList
 4   v = 0; addSol = ∅; freeBoxes = ∅; moreBoxList = ∅
 5   set flag as Preference
 6   while lVol + uVol > sBoxVol * 2
 7        v = (lVol + uVol)/2
 8        Solve SelectMoreBoxes(v) using a commercial solver,
          let moreBoxList be the solution
 9        (addSol, freeBoxes) = 3D-MKP(CC, moreBoxList, flag)
10        if freeBoxes = ∅
11             lVol = v
12        else
13             uVol = v
14   if addSol = ∅
15        return (BoxList, ∅)
16   return (moreBoxList, addSol)
```

An integer programming model, **SelectMoreBoxes**(v), is solved using a commercial software in each iteration of the binary search.

$$z = \max \sum_{i=1}^{N} p_i y_i \tag{1}$$

$$\text{s.t.} \sum_{i=1}^{N} v_i y_i \leq v \tag{2}$$

$$y_i \geq b_i, \ \forall \, i = 1, \ldots, N \text{ and integer} \tag{3}$$

Given a target volume v, constraint (2) requires that the total volume of all boxes in the new order should be smaller than v. The optimal solution to this IP model is the one with the maximum total box preference. The packing patterns for *moreBoxList* are denoted as *addSol*. Whenever a new set of boxes is generated, the procedure 3D-MKP is called to sequentially load them into the container combination *cscc*.

We develop a heuristic method to solve the MCLPP-Delete in the third step. Firstly, a large container combination that is cheaper than *cscc* found in the first step will be generated. Next we try to load a subset of the original order with a high total preference into this container combination, which forms a new order. Packing patterns for the new order will be generated which are denoted as *deleteSol*. The remaining boxes that are excluded from the order are stored as *freeBoxes*. Actually, we may get a solution to the MCLPP-Delete when we are searching for the *cscc* in the first step. In Step 1 if the initial container combination cannot hold all boxes in the original order, a loop will be called to search for larger container combinations until one is found that all boxes can be packed. The penultimate container combination found in the loop can be used as the one we are looking for in this procedure.

After the second and the third step, we will get two loading plans *addSol* and *deleteSol* together with two new orders. They are both candidate solutions to the MCLPP. According to the problem definition, we choose the one with a higher ratio of the total preference of boxes to the total cost of containers (RPC) as the final solution to the MCLPP.

Algorithm 2. Three-Step Heuristic Approach

TS-heuristic()

1   *BoxList* = all boxes in the original customer order
2   *CList* = all available types of containers
3   Invoke FINDCSCC such that $(cscc, origSol)$ = FINDCSCC$(CList, BoxList)$
4   Invoke ADDBOX such that $(moreBoxList, addSol)$ = ADDBOX$(BoxList, cscc)$
5   Invoke DELETEBOX such that $(freeBoxes, deleteSol)$ = DELETEBOX$(BoxList, cscc)$
6   **if** $addSol = \emptyset$
7   *moreBoxList* = *BoxList*
8   *addSol* = *origSol*
9   **if** the RPC *addSol* is smaller than or equal to that of *deleteSol*
10   *bestSol* = *addSol*
11   **else**
12   *bestSol* = *deleteSol*

---

3   Computational Experiments

Our three step search method was implemented as a sequential algorithm in Java and no multi-threading was explicitly used. The experiments are performed on an Intel Xeon E5430 with a 2.66GHz (Quad Core) CPU and 8 GB RAM running the CentOS 5 Linux operating system. The commercial integer linear programming solver used is ILog CPlex 11.0 at its default settings.

Table 1 Experiment results for the MCLPP

BRPC	MTC3	MTC4	MTC5	MTC6	MTC7	MTC8	MTC9
Instance 1	0.057	0.039	0.028	0.031	0.025	0.019	0.016
Instance 2	0.058	0.034	0.031	0.026	0.024	0.018	0.016
Instance 3	0.056	0.036	0.035	0.030	0.025	0.017	0.016
Instance 4	0.061	0.035	0.029	0.029	0.027	0.019	0.017
Instance 5	0.042	0.042	0.031	0.028	0.024	0.018	0.016
Instance 6	0.046	0.034	0.031	0.030	0.026	0.017	0.016
Instance 7	0.047	0.043	0.026	0.024	0.025	0.017	0.016
Instance 8	0.055	0.034	0.027	0.024	0.024	0.019	0.015
Instance 9	0.059	0.039	0.031	0.030	0.026	0.019	0.016
Instance 10	0.044	0.038	0.036	0.029	0.023	0.019	0.016
Instance 11	0.047	0.037	0.028	0.028	0.026	0.016	0.016
Instance 12	0.046	0.037	0.033	0.025	0.024	0.019	0.016
Instance 13	0.052	0.041	0.028	0.027	0.025	0.021	0.016
Instance 14	0.042	0.048	0.030	0.024	0.024	0.016	0.016
Instance 15	0.059	0.038	0.035	0.023	0.027	0.017	0.015
Instance 16	0.063	0.042	0.040	0.034	0.024	0.023	0.025
Instance 17	0.054	0.045	0.036	0.026	0.029	0.024	0.019
Instance 18	0.063	0.041	0.036	0.027	0.029	0.028	0.019
Instance 19	0.059	0.033	0.039	0.029	0.024	0.027	0.027
Instance 20	0.050	0.040	0.039	0.031	0.027	0.021	0.024
Instance 21	0.061	0.043	0.041	0.025	0.029	0.019	0.023
Instance 22	0.045	0.042	0.037	0.025	0.027	0.025	0.027
Instance 23	0.064	0.041	0.041	0.025	0.026	0.023	0.018
Instance 24	0.049	0.042	0.040	0.031	0.028	0.024	0.027
Instance 25	0.059	0.035	0.039	0.032	0.025	0.021	0.027
Instance 26	0.064	0.044	0.030	0.031	0.031	0.026	0.020
Instance 27	0.050	0.039	0.042	0.033	0.024	0.022	0.022
Instance 28	0.055	0.039	0.030	0.029	0.024	0.025	0.024
Instance 29	0.056	0.038	0.041	0.026	0.028	0.028	0.020
Instance 30	0.063	0.043	0.030	0.033	0.030	0.027	0.021
Instance 31	0.064	0.036	0.032	0.031	0.022	0.028	0.027
Instance 32	0.061	0.043	0.031	0.033	0.029	0.023	0.024
Instance 33	0.050	0.038	0.038	0.030	0.029	0.026	0.024
Instance 34	0.066	0.040	0.032	0.030	0.021	0.020	0.019
Instance 35	0.062	0.044	0.038	0.027	0.023	0.019	0.026
Instance 36	0.065	0.045	0.032	0.034	0.023	0.020	0.020
Instance 37	0.054	0.046	0.030	0.029	0.024	0.022	0.019
Instance 38	0.052	0.052	0.034	0.026	0.023	0.021	0.018
Instance 39	0.062	0.047	0.030	0.031	0.023	0.020	0.018
Instance 40	0.074	0.051	0.034	0.039	0.025	0.019	0.019
Instance 41	0.062	0.048	0.039	0.031	0.025	0.017	0.018
Instance 42	0.068	0.055	0.032	0.033	0.026	0.018	0.020
Instance 43	0.065	0.061	0.031	0.035	0.021	0.021	0.019
Instance 44	0.064	0.051	0.039	0.030	0.028	0.020	0.019
Instance 45	0.043	0.048	0.034	0.029	0.020	0.023	0.021
Instance 46	0.067	0.050	0.032	0.035	0.027	0.022	0.018
Instance 47	0.073	0.039	0.034	0.031	0.025	0.020	0.019
Instance 48	0.080	0.056	0.036	0.028	0.025	0.023	0.016
Instance 49	0.071	0.037	0.028	0.032	0.025	0.019	0.020
Instance 50	0.062	0.052	0.034	0.034	0.026	0.019	0.020

Our testing data is based on the 7 data sets created by Che et al. [1]. They are labeled *MTC3, MTC4, ..., MTC9* and there are 50 instances in each *MTC* set. The numbers after "MTC" in the labels indicate the number of different types of boxes in each instance. Three container types with dimensions equivalent to shipping containers of standards sizes (namely the 20', 40', and 40' high cube containers) are used in the *MTC instances*. We modify these test instances by attaching a preference indicator to each type of boxes. The preference is calculated by equation (4), where b_i is the number of type i boxes in the customer order.

$$p_i = b_i / \sum_{i=1}^{N} b_i \qquad (4)$$

Recall that the ratio of the total preference of boxes to the total cost of containers (RPC) measures the quality of a solution (RPC). A larger ratio indicates a better solution. Since there is no prior literature on the multiple container loading problem with preference, we report the RPC of the best solution (*BRPC*) for each instance in Table 1. They can serve as benchmark for future research.

4   Conclusions

This paper presented the Multiple Container Loading Problem with Preference (MCLPP), motivated by a practical application in the logistics department of a large audio equipment manufacturer.

We developed a three-step heuristic algorithm and create test instances for the MCLPP. Experimental results show that our method can provide high quality solutions to the MCLPP and they satisfy the need of the audio equipment manufacturer.

References

1. Che, C.H., Huang, W.L., Lim, A., et al.: The multiple container loading cost minimization problem. European Journal of Operational Research 214(3), 501–511 (2011)
2. Eley, M.: A bottleneck assignment approach to the multiple container loading problem. OR Spectrum 25(1), 45–60 (2003)
3. Wei, L.J., Zhu, W.B., Lim, A.: A goal-driven prototype column generation strategy for the multiple container loading cost minimization problem. Working paper (2012)
4. Zhu, W.B., Huang, W.L., Lim, A.: A prototype column generation strategy for the multiple container loading problem. European Journal of Operational Research 223(1), 27–39 (2012)

Author Index

Abu Jaradeh, Lina   61
Acosta, Gerardo   183
Adham, Ahmed Mohammed   169
Ahmad, Robiah   169
Al-Areqi, Samih   19
Al-Juaidy, Heba   61
Anisseh, Mohammad   97
Ansari, Shifta   47
Arai, Yuya   107

Berzins, Gundars   123
Borri, Dino   3
Borzemski, Leszek   149
Boudraa, Bachir   205
Boudraa, Malika   205

Camarda, Domenico   3
Charles, Darryl   11
Chen, Xuzhi   189

Danovich, Vadim   123
Deb, Kaushik   213

Fu, Meng-Hsuan   83

Ghommam, Jawhar   183

He, Wei   189
Hetmaniok, Edyta   157, 175
Huang, Yangyang   77

Imada, Miyuki   131
Islam, S.M. Towhidul   213
Ito, Takayuki   27, 55

Jo, Kang-Hyun   213

Kamińska-Chuchmała, Anna   149
Klein, Michel C.A.   163
Komatani, Kazunori   69
Kuo, Yau-Hwang   83

Lee, Kuan-Rong   83
Lim, Andrew   219
Ling, Weixin   41
Liu, Wo-Chen   83
Lopez, Alvaro   115

Markič, Štefan   197
Maruyama, Satomi   55
McNeill, Michael   11
McSherry, David   11
Meng, Zhijun   189
Merazi Meksen, Thouraya   205
Mercer, Robert E.   47
Modena, Gabriele   163
Mohd-Ghazali, Normah   169
Mori, Shojiro   69
Mouhoub, Malek   35

Nagai, Akihiko   55
Nakagawa, Hiroyuki   107
Nechval, Konstantin   123
Nechval, Nicholas   123
Nishikawa, Chika   55

Obeid, Nadim   19, 61
Ohsuga, Akihiko   107
Ohta, Masakatsu   131
Okuno, Hiroshi G.   77

Orihara, Ryohei   107
Otsuka, Takanobu   27
Otsuka, Takuma   77

Paul, Richard   11
Picos, Rodrigo   183
Potter, Walter D.   41

Qutaishat, Duha   61

Rogan, Peter   47
Rozenfeld, Alejandro   183

Sadaoui, Samira   35
San Segundo, Pablo   115
Sato, Satoshi   69
Shahraki, Mohammad Reza   97
Shil, Shubhashis Kumar   35
Słota, Damian   157, 175

Stankovski, Vlado   197
Sultana, Kazi Zakia   213
Szlávik, Zoltán   143

Tahara, Yasuyuki   107
Tapia, Cristobal   115
Tian, Tian   219
Tsuboi, Tatsunosuke   27

Wang, Kaipeng   189
Warmerdam, Vincent Damian   143
Wituła, Roman   175
Wu, Jiansheng   91

Yi, Yang   91

Zhang, Haiyi   91
Zhu, Wenbin   219
Zielonka, Adam   157, 175

Printed in the United States
By Bookmasters